Specifying Buildings

A design management perspective

Stephen Emmitt
and
David T. Yeomans

OXFORD AUCKLAND BOSTON JOHANNESBURG MELBOURNE NEW DELHI

Butterworth-Heinemann
Linacre House, Jordan Hill, Oxford OX2 8DP
225 Wildwood Avenue, Woburn, MA 01801-2041

A division of Reed Educational and Professional Publishing Ltd

First published 2001
Transferred to digital printing 2004

British Library Cataloguing in Publication Data
A catalogue record for this book is available from the British Library

Library of Congress Cataloguing in Publication Data
A catalogue record for this book is available from the Library of Congress

ISBN 0 7506 4849 X

For information on all Butterworth-Heinemann publications visit our website at www.bh.com

Typeset by Avocet Typeset, Brill, Aylesbur

Contents

List of illustrations

Preface

The correct specification of buildings is essential if parameters such as quality, time, value, buildability, maintenance and durability are to be met: yet, guidance on how to select building products from the vast quantity of information available is in short supply. Existing literature is concerned with the physical writing of the specification, not the selection process that precedes it or the pressures faced by the specifier to change specified products as the project passes through tendering and site operations. Furthermore, there is little guidance available to the manager of the professional office on how to manage the specification process. We have tried to address some of these shortcomings by focusing on an important, but little researched, aspect of design practice and building construction – the process of product selection and building specification.

The scope of this book is relatively straightforward. We have combined the physical act of specification writing with the act of product selection and the management of the process. In particular, we consider how new products are taken up by an industry where specifiers are anxious to avoid unnecessary exposure to risk and describe the mechanisms used by design organisations to filter the information coming to them through a series of formally and informally established 'gates'. Case studies based on the observed behaviour of specifiers are used to illustrate issues from which lessons may be drawn. These observations have their roots in doctoral research, which applied the diffusion of innovations model to the building industry in an attempt to study the uptake of new building products. Although based on the observation of specifiers in an architectural office, our experience suggests that these studies are applicable to other design professionals, such as building surveyors, who carry out similar tasks. We have also broadened the work through additional case study material based on interviews with specifiers working in a wide range of design offices engaged in both new build and refurbishment projects. We attempt to show specification as a much richer and more interesting area than the books devoted entirely to specification writing.

Our book is aimed at students and young practitioners as well as the principals/managers of design offices who need to manage the specification process to ensure a quality artefact. In addition to advice on specification writing, we have attempted to address the related issues of product selection and the management of the specification process, which should also be of interest to those attempting to sell their new products to specifiers. Because there are differences in specification practice between different countries, we have attempted to deal with the underlying principles in a generic manner so that the book is of value to specifiers wherever they happen to practise.

Stephen Emmitt and David T. Yeomans

Acknowledgements

This book would not have been possible without the help of many specifiers (architects, engineers, technologists and surveyors), design managers, contractors, building product manufacturers and their trade representatives who were kind enough to talk to us and/or let us observe them in action. We are extremely grateful.

We would also like to thank the Manchester Society of Architects (MSA) who saw potential in our work back in the early nineties and for the award of the MSA Scholarship (to S.E.), which, in a roundabout way, led to this book. We would also like to acknowledge the RIBA's help through a research award that allowed us to look at the gatekeeping process in greater detail. Thanks are also due to Ian Dickenson at Leeds Metropolitan University for turning our sketches into artwork. We also gratefully acknowledge the unknown reviewers and the staff at Butterworth-Heinemann who helped us keep to the specification.

Stephen Emmitt and David T. Yeomans

Specifications in context

This introductory chapter starts by looking at specifiers and the specification process before turning to the written specification. The historical background of specifications is followed by an overview of the range of architectural components to be specified. The chapter concludes by looking at the contextual issues facing all specifiers, regardless of the type of office in which they work or their location.

Specifiers and specification

For many years, concern has been expressed about the quality of buildings and the quality of the processes that lead to the production of our built environment. Much research, debate and attention have been focused on procurement options, cost, time and value. Very little attention has been given to the one document that defines quality standards for both materials and workmanship, the written specification. Of all the information provided and used in the process of building, it is the written specification that describes the quality of the final artefact. Given that many defects in buildings can be traced back to inadequacies in workmanship and/or materials, or inadequate information, one could be excused for asking why this central document and the process that leads to its production have been neglected.

Information is central to the design and construction process. Drawings, models, schedules and bills of quantities cannot convey the whole message; they have to be supplemented with descriptive information. On very small projects, this information is often provided in the form of notes on drawings, but for the majority of projects, the descriptive information is extensive and needs to be accommodated in separate documentation known as a 'specification'. Designers specify their intentions, and these (along with other contractual information) are interpreted by the contractor and translated into a physical building. The quality of a building is determined by a combination of factors: the design, selection and specification of building materials and products;

the accuracy of contractual information; and the ability of the contractor and sub-contractors to interpret information in the form of lines, words and figures into a completed building. It is the written specification, not the drawings, in which quality standards are laid down to be achieved by the operatives on-site. Thus, if we are to achieve and improve the quality of the buildings around us, a good starting point would be to understand and manage the specification process in its entirety.

The word specification is used in two different ways in the building industry. On the one hand, it is a term used to describe the selection of a product by a specifier, and on the other hand, it refers to a physical document that contains a written description of standards of workmanship and the performance of materials and products. To avoid confusion, *specification* is used to describe the process of selection of building products and *written specification* for a description of the document. The term *specifier* is used to cover all those involved in product selection, regardless of their background, be they architect, engineer, technologist or contractor.

The specifiers

Both the quality and the long-term durability of a building depend upon the selection of suitable building products *and* the manner in which they are assembled on site. The selection of building products or, more specifically, the selection of the correct building products for a specific purpose, within the limits of cost and time, is an important task for the specifier. It is, however, a process that is difficult to observe in its entirety and one difficult to describe since it relies on the specifier's tacit knowledge. Building product selection is an essential and familiar process to practitioners, yet, in spite of the frequency with which products are specified and the importance of the resulting decision, the act of specification has rarely been observed. It is implicit in design training but is rarely taught as a distinct skill, and furthermore (and perhaps surprising), research into the field is sparse. Traditionally, architects have been the major specifiers, but with the introduction of new procurement methods and the growth of other specialists, there has been a move to specification by a wide variety of building professionals. Although architects' responsibility for building product selection has declined in recent years according to the Barbour report (Barbour Index, 1993), they are still the most influential and important 'specifiers' of products in the British building industry because of their influence at the design stage. Pawley (1990:5) observed that:

> This role as selector of materials and components is now the core of the status of the architect in the construction industry. It ensures that a steady supply of component and materials advertising is directed towards architects; as long as they

> continue to choose key components and finishes they will remain in effect 'licensed specifiers', with powers and responsibilities similar to those of doctors of medicine who prescribe drugs.

This influence over the majority of building product specification is well recognised by building product manufacturers who bombard architects' offices with their trade literature in the hope that specifiers will specify their products in preference to those of their competitors.

As long ago as 1933, the architect Chermayeff, in an article entitled 'New Materials and New Methods' (Chermayeff, 1933), emphasised the importance of specifying the correct material for a given situation when he said that:

> It is essential to select for a specific purpose within the defined cost, the most adequate material and method; that is to say, that material which best solves the problems of purpose, money and time.

This still holds true today. The challenge then for the specifier is to ensure correct specification within the time available. The task for the office manager is to provide the conditions that make this possible.

Recently, a number of new pressures have been put on the specifier. One of the points noted by Latham (1994) in his much-cited report into efficiency in the UK industry was a concern about over-specification, i.e. the specification of components of a higher standard than necessary. Both the Latham report and the later Egan report (Egan, 1998) were concerned with reducing costs and improving efficiency in the UK construction industry. There has also been a growing interest in the selection of 'environmentally friendly' products, which implies a change in specification practice. While this might have led to a reappraisal of the components being used and assembled and a switch to new products, there has been an increase in risk management techniques by specifiers' offices in an attempt to reduce potential claims for negligent selection of building products. This may have resulted in the tendency to limit the number of changes to products used previously and also a reluctance to use products that are new to the office. The correct specification of building products is a complex task, with the specifier subject to a number of sometimes contradictory pressures.

The specification process

Before building products can be considered for their suitability for a given situation, the performance requirements for a particular project must be established (Taylor, 1994). Each building project is unique in situation, more often than not unique in design, and frequently unique in the individual members that make up the temporary project team responsible for the building's design and assembly. Designers need to ensure correct specification, ensure that the specification is transmitted to the building site effectively, and check that work on site is to the specified quality standards. The process can be divided into three key components (Figure 1.1):

- decision-making;
- communication;
- control.

In order to operate efficiently, specifiers will require rapid access to current and relevant information; they must possess adequate technical knowledge, have the ability to convey instructions clearly and unambiguously to the builder, and (subject to the appropriate contractual arrangement) must ensure that products are not substituted nor performance standards compromised without their knowledge and approval. As demonstrated in later chapters of this book, the specification process is complex, difficult to observe in its entirety, and demanding in terms of its management. It may come as a surprise to some readers that despite the amount of time designers devote to the process of specification, there has been very little research into it. That which has been published has found the process to be both complex and subtle as well as being influenced by a wide range of people at different times in the life of a project. These are issues taken up later: first, we need to look at the framework in which the specifier has to operate.

A framework

The manner in which a building is designed and then built is rarely a neat and ordered process, there are frequent changes, re-designs and re-programming as the project moves from the initial idea to a finished building. A number of conceptual project plans exist that aim to guide the designer through a project; the most widely known of which is the RIBA's 'Plan of Work', which implicitly divides the design process into separate stages. Although the plan of work has been criticised because it does not allow for feedback loops, which would allow new information to be incorporated in the ongoing decision sequences, it continues to be used by practitioners as a guide to organisation and resourcing of projects, as well as a guide for fee invoicing. Research has found that designers do not adhere rigidly to the plan, but work closely to it (e.g. Mackinder, 1980). From the perspective of the specifier, there are a number of distinct phases in the specification process during which the individual will be engaged in activities that differ from the preceding stage. The relationship between these stages

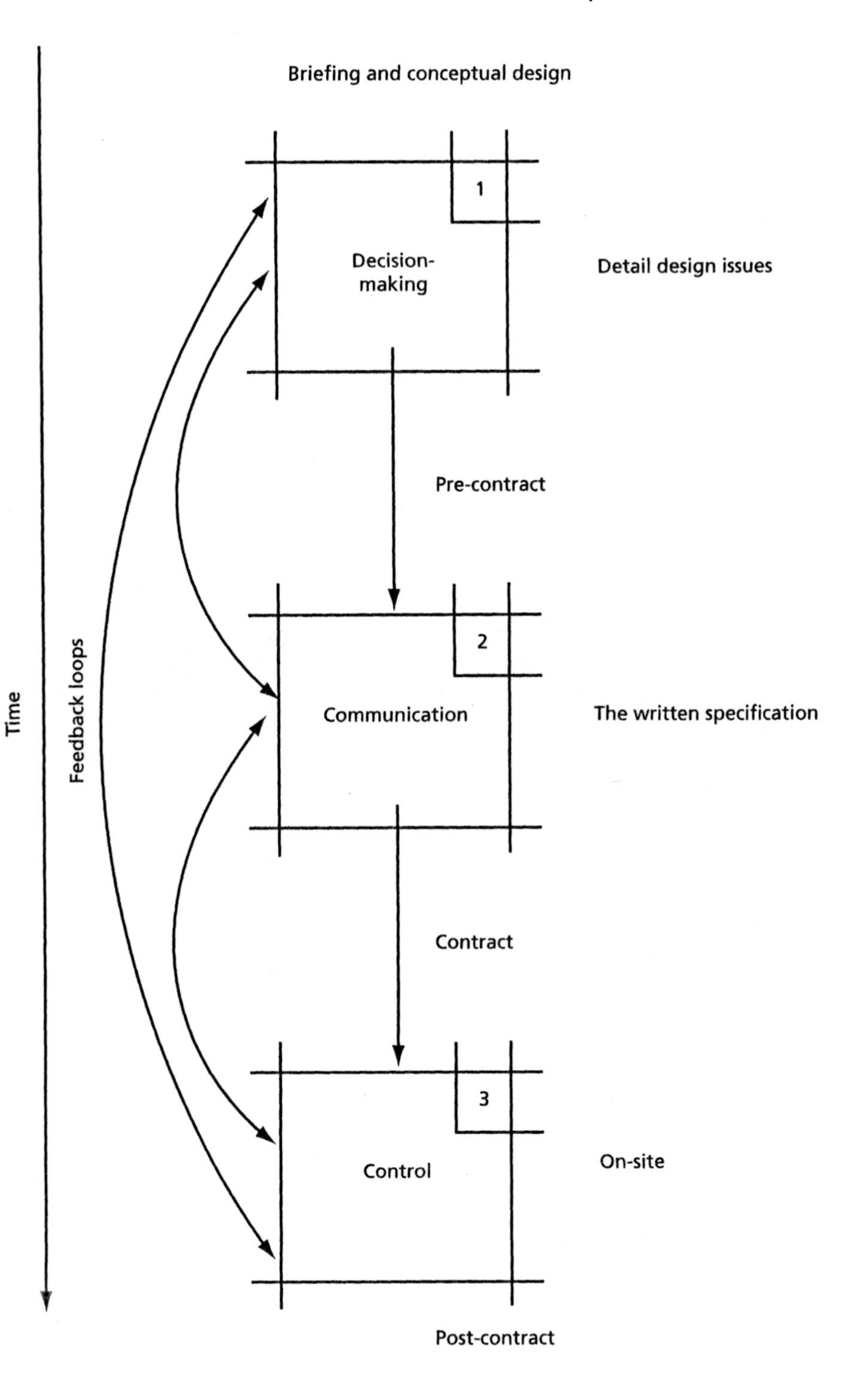

Fig. 1.1 The specification process

and building product selection is shown in Figure 1.2 and may be conveniently divided into three distinct phases, namely Conceptual Design, Detail Design and Production, although it is recognised that in practice, there may be some overlap between these. American readers will note that the AIA Handbook (AIA, 1988) has five stages in an architectural project: schematic design, design development, construction documents, bidding or negotiation, and construction. For simplicity, we have grouped the latter three stages under Production.

1 Phase 1, Conceptual Design (RIBA stages A, B, C and D). It is at the briefing stage where performance parameters for the intended building should be agreed and confirmed in writing to form the design brief. The designer should also be considering the life cycle of the building, its maintenance and final disposal strategy at this early stage, but there is little evidence to show how this is seriously considered in design except when issues of serviceability and durability are involved. Sustainability is a relatively new idea, an innovation as a design concept, and while there may be considerable lip service paid to the idea, it is questionable whether it has had much influence on actual design processes (Emmitt, 2000). Following the production of a working brief, a feasibility study and sketch designs are produced (plans and elevations) for client approval prior to submission for town planning consent. Here, it is common to confirm generic materials to be used on the external faces of the building (e.g. 'red facing brickwork'; 'profiled metal cladding, steel blue'), and therefore, some material selection is occurring early in the process, coincident with the architect's overall design vision and the need to obtain consents from the client and the town planners. The point is that these initial decisions about major materials will influence the decisions made in phase 2.
2 Phase 2, Detail Design (RIBA stages E, F, G and H). This involves the production of a large number of drawings and schedules that will enable the production of the bills of quantities, enable selected contractors to negotiate or bid a contract price and enable the chosen contractor to build the building. During the production of the detail drawings (comprising assembly drawings, component drawings and schedules), the designer will be making decisions that relate to specific building product selection. Detailed input from consultants such as structural engineers, quantity surveyors and also the client may influence specification decisions during this phase. Furthermore, it is during this phase that conformity with building codes, regulations and standards must be ensured. At least, that is the conventional picture in the UK because it is in this phase that there may be different organisational arrangements. For example, there is an increasing use of performance specifications by architects, which leaves the choice of actual product to the contractor and sub-contractors. In France, the activities involved in these first two phases will actually be carried out by different kinds of office, the architects who produce the design and the *bureau*

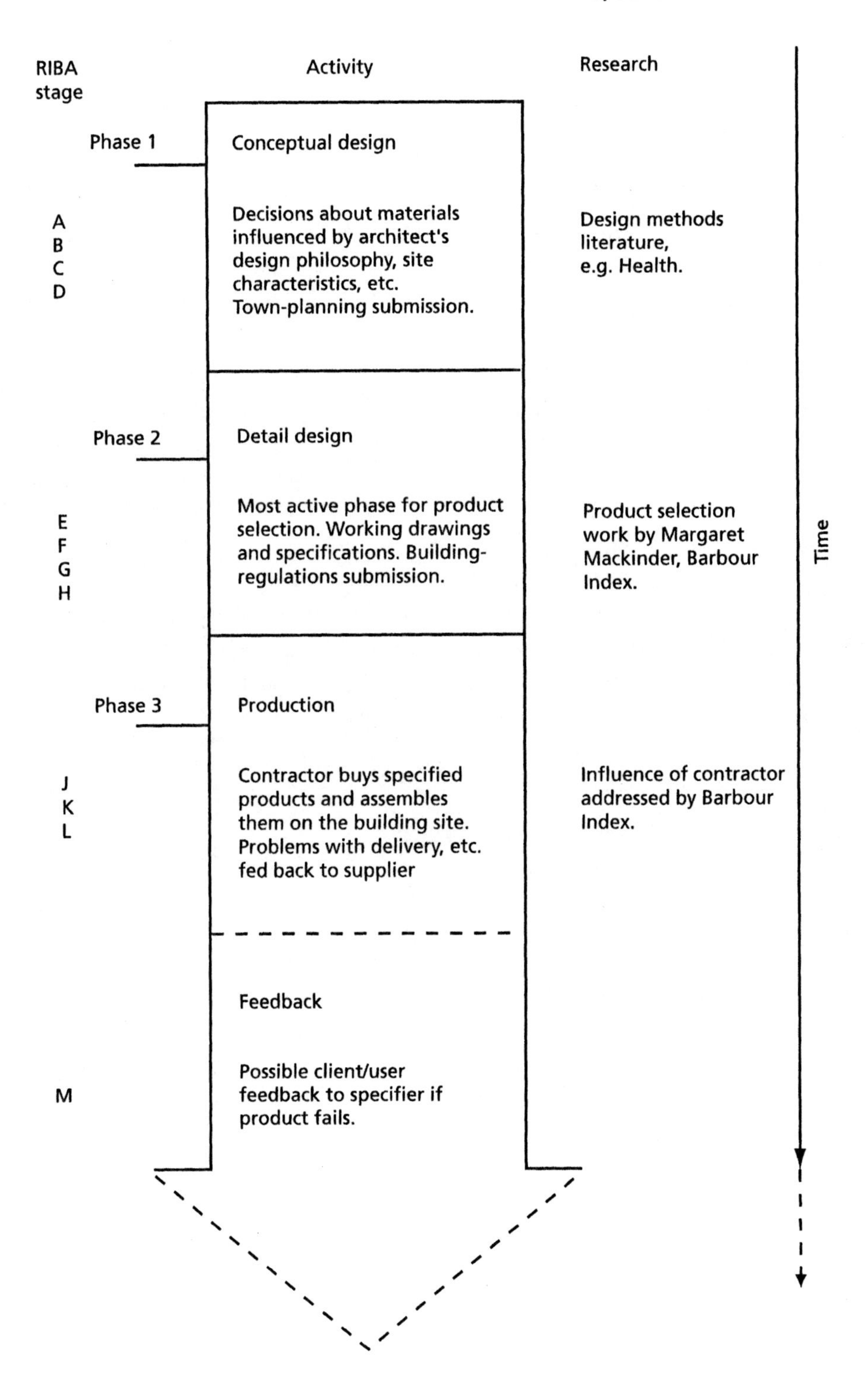

Fig. 1.2 Product selection and the design process

d'etudes, who produce the production drawings. Indeed, this separation of responsibilities between offices can also be found within larger offices elsewhere, where the detailing and specification writing is carried out by someone other than the conceptual designer.

3 Phase 3, Production (RIBA stages J, K, L and M). This is essentially a phase concerned with the realisation of the design in the form of a finished product, the translation of abstract ideas to a physical artefact, the building. Although all specification decisions should have been confirmed by the end of stage F, there may be pressure to change the level of performance or specified products to cheaper ones. This may be because the cost of the building is over budget. But pressure to change may also come from the contractor and sub-contractors because of supply difficulties, to improve buildability on site and/or to reduce product costs. It is also possible for changes to be imposed by client bodies and external control agencies. The way in which this occurs has been illustrated by Andrews and Taylor (1982) who give an account of how major changes affecting the appearance of the building may be imposed upon the architect by external factors quite beyond his control. More recently, a TV series and accompanying book (Sabbagh, 1989) examined the construction of a New York skyscraper, which provided an insight into how the working of designers continues into the construction phase, but such 'natural' histories of the design process are regrettably few (Yeomans, 1982).

Detail design, the most active phase

Although numerous decisions are being made throughout the design process that will inform the writing of the specification, the most active stage of any project in terms of specification decision-making is the detail design phase. This phase involves the production of a large number of drawings and schedules, which will both enable the production of the bills of quantities and be used by the contractor to construct the building. When the project enters the detailed design stage, the decision-making process is different from the conceptual stage for a number of reasons. First, the designer will be constrained to a certain extent by the conceptual design and any performance parameters that have been established; thus, the problems may be better defined than in the earlier stages. Second, a number of individuals will be involved in the development of the details, other consultants, manufacturers, contractors, cost consultants and project managers, and this will influence decisions that lead to the writing of the specification.

During the detail phase, the designer will be trying to finalise the construction details, so the relationship between detailing, product selection and the writing of the specification is paramount. Because the resulting information will be used by the contractor to assemble the building, any errors or discrepancies between the documents may well lead to disputes, litigation and arbitration. At the tendering stage and throughout the construction phase, there may be pressure to change materials and products (for a variety of reasons), and so the decision-making process may well extend into the construction phase. It is during phase 2 that specifiers will be actively looking for solutions to their particular detailing problems and therefore may be more receptive to information about building products that are new to them. At the end of phase 2, the specification must be complete, so decisions about how to detail particular junctions, and hence what products and/or performance levels are required, must be determined and confirmed in the written specification.

The written specification

Specifications are written documents that describe the requirements to which the service or product has to conform, i.e. its defined quality. Like drawings, specifications vary in their size, layout and complexity. In all but the smallest of design offices, it is common for specifications to be written by someone other than the designer; thus, communication between designer and specification writer is particularly important. The majority of designers are visually orientated people whose skills are best employed in the conceptual and detailed design phases. Therefore, few have time to be involved in the physical writing of the document: this task is usually undertaken by a technologist or specification writer, someone with better technical and managerial skills than many designers possess. Specification writers require an appreciation of the designer's intention and the ability to write technical documents clearly, concisely and accurately. They also need to be able to cross-reference items without repetition. Standard formats form a useful template for specifiers and help to ensure a degree of consistency as well as saving time. In the UK, the National Building Specification (NBS) is widely used because it helps to save time in this way and is familiar to other parties to the design and assembly process (see Chapter 5).

It is usual practice to use the term 'specification' in the singular, which is a little misleading. In practice, the work to be carried out will be described in specifications written by the different specialists involved in the project. Even on the most simple building schemes, the engineer will write the specification for the structural elements such as foundations, the architect will be concerned with materials and finishes, and there will be an electrical and mechanical specification, possibly a specification for the landscaping works, and one for highways work. This collection of multi-authored information is known as 'the specification'.

Specification types and uses

Responsibility for the written specification is the architect's, although in Britain, it is not uncommon for the quantity surveyor to contribute to its preparation. Many quantity surveyors complain about their input to the written specification, forced to do so because the document is sent to them containing errors and omissions. This may be considered as careless or just lazy practice on the designer's behalf, although the situation usually arises because insufficient time has been allocated to the task, i.e. it is a management problem.

There are two ways in which building work may be specified, namely through the use of prescriptive specification or by performance specification (see Chapter 2). Prescriptive specification is the most popular method where the designer produces the design requirements and specifies in detail the materials to be used (listing proprietary products), methods and standard of workmanship required. Performance specification is where the designer describes the material and workmanship attributes required, leaving the decision about specific products and standards of workmanship to the contractor. The advantages and disadvantages of using one over the other are discussed in the following chapter. However, it is common for both prescriptive and performance methods to be used in the same document to cover different building elements.

The written specification will be used for a number of quite different tasks by people from different backgrounds. It will be developed and read in conjunction with drawings and schedules. Requirements and standards confirmed during the briefing process will form a series of parameters from which to develop the design and the specification during the pre-contract period, will be transmitted to site, and will form a record of the work for future reference. Good specification writers are aware of the different uses to which the document will be put and the different backgrounds of those likely to read it (see Figure 1.1). More specifically:

During the pre-contract phase, the document will be:

- developed from, and be central to, the briefing process;
- read by the cost consultant to prepare cost estimates and the bills of quantities;
- read by the contractor's estimators to prepare the tender price.

During the contract, the document will be:

- read by the contractor's agent on site;
- read by site operatives;
- read by the clerk of works and/or resident architect to check the work is proceeding in accordance with the contract documentation.

Post-contract, the document will be:

- used as design record of materials used and set standards;
- used as a source of evidence in disputes;
- used as a source of information for maintenance, facilities management and recovery management;
- analysed for feedback into the master specification and office procedures.

An historical note

Writing in 1985, Jack Bowyer noted that specifications had changed little in essence from those available at the start of the 20th century. At the start of the 21st century, a similar observation can be made: the documentation has become more extensive than it was in the past, but its purpose of laying down standards to be met remains the same. Likewise, books providing guidance to students and practitioners have changed little in their main message. What does tend to change is the fluctuating fashion for the use of performance over prescriptive methods and vice versa. Both approaches to specification are considered in this book, the choice of one method over another being a matter for individual design organisations and their specifiers. At the time of writing this book, performance specifications are coming back into fashion. However, adopting a performance approach does not eliminate the task of selecting proprietary products: it merely passes the decision-making process down the line to the contractor and/or the sub-contractors who have to make their selection from a range limited by the designer's performance parameters. Thus, the decision-making process observed and described later in the book is appropriate to both prescriptive and performance specifications.

Many of the developments in specification have occurred in the past 50 years or so (see Cox, 1994). In the United States of America, the Construction Specifications Institute (CSI) was founded in 1948 to serve the interests of specifiers and manufacturers. In 1963, the CSI and Construction Specifications Canada (CSC) worked together to implement standards and published the 16 division Masterlist of specification sections in 1963 – now the 'Masterformat'. The CSI *Manual of Practice* was first published in 1967 and revised and updated on a regular basis. Australian specifiers have seen the development and refinement of the NATSPEC system. In the UK, no specification writing standards existed until the publication in 1987 of the Common Arrangement of Works Sections (CAWS): until this time, most specifications had been arranged under the same headings as the Bills of Quantities (Cox, 1994).

For such an important aspect of the designer's (and engineer's) job, it is a little surprising to find that there is very little published work in the field. Textbooks have restricted themselves to guidance on the act of writing the specification with little or no recognition of the selection process that is an integral part of specification writing. Published research into the process is sparse.

In 1980, Margaret Mackinder looked at how architects made decisions about building materials and products, and she found that apart from work relating to information flow and retrieval, plus a small amount of work on specification writing, no work on how professionals actually select materials had been carried out. The situation has changed little since 1980. In 1981, the *Architects' Journal* reported a telephone survey carried out by a commercial market research company, Walton Markham Associates, and a short report by Moore (1987) looked at contractors' influence on specification decisions. More recently, the Barbour Index, a commercial supplier of information for specifiers, published a series of reports (Barbour Index, 1993, 1994, 1995, 1996).

A natural question to ask is: to what extent do students learn the art of specification during their studies? Mackinder (1980) looked at the extent to which schools of architecture taught the selection of building products and found that they did not. Schools teach what they think of as design, although it could be argued that material and product selection is an important aspect of this, but it is not much considered. This has been criticised by, for example, Crosbie (1995) and Antoniades (1992:222), who believed the 'major weakness' of architectural design in schools of architecture to be the lack of attention given to the importance of materials and building technology. The effect is that the Schools teach what has been defined as phase 1 (RIBA stages A–D) leaving phases 2 and 3 to be learned in practice. This appears to be a common model in both Europe and the US, one exception being Australia where the RAIA are explicit in the requirement for members to demonstrate competency in specifications (Gelder, 1995). The habits of product selection are therefore passed on during a young designer's 'apprenticeship' within the design office and so will be strongly influenced by practice procedures and the influence of more experienced members of the organisation. Thus, the tendency for specifiers to select the same products used by their colleagues and their office is strong. Without any discussion on specification practice during education, architects will be ill equipped to take a detached view of such procedures; this may well result in the perpetuation of obsolete behaviour that has become nothing more than a ritual. We should note that the recent introduction of the undergraduate degrees in Architectural Technology during the 1990s in the UK should help in this regard. These degrees incorporate specification skills as part of the technologist's competency, and, combined with education in design management techniques, these graduates should be well equipped to challenge existing practices (Emmitt, 2001a).

Parallels can be seen in the prescription of drugs by medical practitioners where prescribing habits are known to form in early clinical practice. Medical schools worldwide are starting to adopt a problem-based approach to learning, so that medical students can develop the skills required to critically evaluate new drugs that come onto the market (MacLeod, 1999). To encourage this approach, the World Health Organisation has produced a teaching aid, *Guide to Good Prescribing* (WHO, 1995), which is designed to help students develop a method for selecting appropriate drugs and be less susceptible to external influences, such as pressure from drug companies.

Architectural components and their selection

At the heart of good architectural design lies the correct selection of materials, components and products that make up a building's assembly: they contribute to the aesthetics and durability of the completed building. In his book, *The Roots of Architectural Invention*, David Leatherbarrow (1993:143) made the observation that:

> Because materials are familiar in experience and unavoidable in construction one might assume this specification is a procedure that can be described simply and clearly; in fact the opposite is true, for it is both a rarely discussed procedure and one that exposes strikingly obscure and indefinite thinking when questioned. Yet this obscurity is unavoidable because material selection is inevitable.

Product selection, above all else, is one of the most important considerations for the long-term durability of the completed building and an area in which building designers should excel. Not only have the selection and specification of building products attracted little attention from researchers, but they are rarely discussed by practitioners, presumably because it is difficult to separate the designer's goals from building materials as entities in their own right (Patterson, 1994). It is a process that is often seen as unglamorous and something that should be carried out as quickly as possible. Architectural literature, especially architectural periodicals, tends to be pre-occupied with the finished appearance of buildings, both inside and out, paying little attention to the process of producing the building. The majority of published literature that has investigated the way in which architects make decisions has concentrated on the design process with emphasis on the resulting form (Rowe, 1987). This body of literature goes back to the 1960s, is commonly referred to as 'Design Methods' literature and is primarily concerned with creative problem-solving (e.g. Thornley, 1963; Heath, 1984). There is a clear distinction in this work between the selection of materials, which may be an inherent part of the designer's design idea, and the selection of specific products to fulfil this vision. The separation of stages can also be seen in some of the larger architectural offices in which the design architect delegates responsibility for the detail design to a technologist. This division has been effectively institutionalised in France with the different types of work carried out by different offices. However, there is evidence that the separation of conceptual from detail design is not liked by practitioners in Britain, a point noted by Mackinder (1980:12):

> Many architects assert that [the design process] 'is' the process of selection, organisation and specification of materials, and refuse to separate the two.

There appears to be a difference between the architect's perception of the process and the actual process, although practices differ in different offices.

Building products

The number of building products that are potentially available for selection by an architect is extensive. There are approximately 20 000 building product manufacturers, many of whom offer more than one product for sale (Edmonds, 1996). Every year, new products are introduced by manufacturers in response to competition, new regulations and changes in architectural fashion. In addition to these 'new' products, there are numerous minor product improvements that are constantly introduced by manufacturers to prolong their product's life on the market. But getting a new idea or product adopted is never easy, as Rogers (1995) has pointed out, and this is especially true of building, where new products and product improvements, like the established products, are dependent upon decision-makers in the building industry for their selection.

Specification of building products is of great importance to the manufacturers and suppliers who do carry out a good deal of commercial research into the adoption of their own, and their direct competitors', products. This research is not in the public domain simply because of its commercial sensitivity and helps to explain why published research is rare. Research into the selection of building products by architects was carried out by Margaret Mackinder (1980), who gathered information about detailed design decisions from diaries filled in by participating architects. She observed that architects frequently used 'short cuts' based on their own experience in order to save time, reporting a strong preference for certain materials and components that they had used previously, drawn from their personal collections of literature. This supports the earlier observations of Goodey and Matthew (1971) and Wade (1977). One-third of Mackinder's sample acknowledged that new materials and methods needed to be monitored but claimed that it was office policy to avoid the use of anything new unless it was unavoidable, preferring to specify familiar products. Mackinder's study found that there had been very little research into how professionals actually selected building products, a situation that has changed little since 1980 (Emmitt, 1997).

Contextual issues

The specifier's office does not work in isolation, it is engaged to provide a service to a client (who may contribute to the process of product selection), and it is involved in communicating the design to the contractor on the building site. Thus, the architectural office, even in the most simple of contractual relationships, may be influenced by contributions from persons outside the architect's social system. These individuals, with different values and goals, are brought together for one particular project, during which both formal and informal communication will take place between them. The manner in which various individuals interact will depend to a large extent upon the nature and size of the proposed building project and the type of contract used to procure the building. Within the contractual structure of each project, there is a constant exchange of instructions and information between the client, the professional team and the contractor; this relationship is known as a *communication network* (Rogers and Kincaid, 1981). Although the communication network is important in the transfer of information, studies carried out in the 1960s (Higgin and Jessop, 1965) indicated that communication between the various individuals was ineffective at all stages in the design and production of buildings with particular mention made of the gulf between the designer and the builder: a situation that, if anything, has become worse over the years.

Time

Regardless of the building type, size and complexity of the design, each project will have some form of time constraint imposed on it. Usually, the client requires a completed building for a particular date, a date that will influence the amount of time allocated to different phases of the project. This imposes time constraints that have to be accommodated into overall programming of resources, thus limiting the amount of time available for producing the requisite information. Adequate time is required to consider appropriate products and set performance standards, co-ordinate information provided by others, write the specification and check the project documentation for consistency and errors. Programming of the specification process is paramount if good-quality information is to be produced. Time constraints also influence the uptake of new products as demonstrated by Mackinder (1980) and Emmitt (1997). When a project had to be completed quickly, there was an increasing tendency to stick to products used on previous projects, thus eliminating the time needed to search for alternatives. The extent to which this action hinders the uptake of new products is explored in later chapters.

Consequences

Every specification decision will have consequences for the durability of the building. There are also consequences for the organisation that specified the products, hopefully good in terms of a job well done and repeat business, but unfortunately, sometimes, the consequences are less welcome with some form of legal action being brought against the design office. Specifiers need to be aware of a wide range of issues that influence their decision-making process, as do the design managers who program and manage the process. Key skills are:

- an ability to make informed decisions;
- the ability to communicate effectively and efficiently;
- an ability to manage the entire specification process.

With the industry becoming increasingly litigious, the specification has taken on a more important role than it had in the past. Like the drawings, the specification is a legal document and will be examined thoroughly should a dispute arise. It is necessary, therefore, for the specifier to have a thorough knowledge of contractual issues. By this, we mean an awareness of procurement options; conditions of contract; drawings; bills of quantities; standards and codes; and an ability to co-ordinate them all in a logical and thorough manner. In practice, this may be more challenging than it might at first appear simply because the specification writer may not have been involved in the preparation of this documentation.

As the building industry has continued to specialise, and hence fragment, the number of ways of getting things done has become more complex. Clients and their professional advisers are now faced with a bewildering choice of procurement routes that continue to evolve. The choice of one method over another is dependent upon the size and type of project as well as on the opinions of those involved. For example, architects either favour traditional methods because they retain a fair degree of control (with contractors favouring contractor-led procurement options for the same reason) or 'design and build' contracts because it reduces their need to become involved with the details. In terms of responsibility for specifying and specification writing, the contractual routes can be divided into traditional (designer-led) and non-traditional routes (contractor-led) in which the designer is merely a sub-contractor. The choice of procurement route will have a direct influence on the contract documentation.

Using traditional forms of contract, the quality of the finished building will be determined by the architect and fellow professionals. With contractor-led procurement, the tendency has been to favour performance specifications so that the choice of product can be determined to assist the contractor (and some would argue, not the client). Cost and delivery are two important criteria for selecting proprietary products that meet the performance parameters. In situations where products are specified by brand name, the contractor retains the right to substitute them, and thus the quality of the finished building and its service life may still be affected by the contractor.

A generic approach

We stated in the preface that this book aims to provide a generic approach to the specification of buildings so that the contents are of value to specifiers wherever they happen to practise. In taking such an approach, we have tried to resist the 'this is how you do it' approach; instead, we have used research findings to illustrate some of the points raised in the book, from which the readers can draw their own conclusions. Because of the generic nature of the material in this book, we have deliberately avoided descriptions of terminology in contract documentation, contractual issues and legal matters, other than matters of general concern. We have also tried to avoid adding examples of standard layouts, partly because of our generic approach and partly because IT-based software packages render such examples largely redundant. Readers will need to refer to the appropriate standards for their particular location, for example, American and British readers may find *Writing Specifications for Construction* by Peter Cox (1994) a useful and thorough guide. We have also tried to address issues relating to responsibility (and hence liability) in a generic manner and would urge readers to seek legal advice for their particular circumstances and contract peculiarities, simply because this varies widely and will change over time as legal precedents are set and subsequently tested.

Readership

We have addressed this book to a relatively wide audience, namely students, young practitioners, design office managers and the manufacturers of building products. So, although we have designed the book to be read from cover to cover, we also appreciate that many readers will want to dip in at various points to suit their particular circumstances. Thus, a small amount of guidance may be useful.

Students will be influenced by their particular course structure and directed by their lecturers, however, we would recommend Chapters 2–5 as essential reading. Young practitioners engaged in product selection and specification writing in practice will find the practical advice given in Chapters 3–5 particularly useful. We would also urge them to read and reflect on the observation of the specifier contained in Chapter 10.

Design managers will be more interested in Chapters 6–11. Building-product manufacturers and their advertising consultants should find Chapters 7–10 particularly useful given the lack of published information in this area. Finally, individuals engaged in research into specification and design management are directed to Chapters 7–11.

Specifying design intent

Designers must communicate their design intent to a wide variety of project participants during the design and the assembly stages of projects. To do this effectively, it is necessary to be specific, through the use of drawings, schedules and specifications. This chapter looks at the relationship between drawings and specifications, the quality levels required, and the differences between prescriptive and performance specifications. The chapter concludes by looking at different approaches to specification and the influences on the process from others party to the project.

Communicating design intent

An essential requirement of the professional design organisation is to be able to produce clear, concise, and accurate information that can be used to assemble the diverse range of materials and components into a building that meets the client's requirements and expectations. With the exception of artisans and the designer-craftsperson, designers work and communicate indirectly (Potter, 1989). Their creative work is expressed in the form of instructions to manufacturers, other consultants, contractors and sub-contractors, usually expressed in the form of drawings and written documents, collectively known as 'production information'. Instructions must be clear, concise, complete, free of errors, meaningful, relevant and timely to those receiving them. Michael Brawne (1992), in his book *From Idea to Building*, makes the point that architectural ideas require considerable effort, discipline and commitment if they are to stand any chance of translation into buildings.

Communication media need very careful consideration – the selection of one over another to satisfy a particular set of circumstances. Drawings and written documents are used to describe and define a construction project. At its best, this project information will be clear and concise and easily understood, and the building contract will proceed to programme and to cost. At its worst, poorly conceived and shoddy project information will lead to confusion, inefficiency, delay, revised work, additional expense, disputes and claims. It is a sad fact that very few projects are perfect: many are flawed by poorly expressed requirements. With the advent of computers and digital information, one could be forgiven for thinking that inadequate information would become a thing of the past, but many within the industry have noted a significant increase in the quantity of information provided and a steady decline in its quality. Unfortunately, computers make it easy to transfer errors from one document to another very quickly. But it would be misleading to blame the technology. The biggest enemy of those trying to produce comprehensive, good-quality information is that precious commodity, time. With increased pressures on time (and fees) has come the requirement to compress the amount of time taken to produce the project information – we should expect mistakes and omissions.

At the heart of this body of information lies the written specification, a document central to determining the quality of the finished building. Yet this document is one of the most poorly considered and misunderstood documents. The written specification (and the process of specification) should not be seen as something that has to be done at the end of the detail design phase. Instead, its relationship to other documents must be recognised so that the specification can be developed during the whole design process as part of a comprehensively managed and co-ordinated set of information.

Co-ordinated project information

Co-ordinated project information (CPI) is a system that categorises drawings and written information (specifications) and is used in British Standards and in the measurement of building works, the Standard Method of Measurement (SMM7). This relates directly to the classification system used in the National Building Specification (NBS). One of the conventions of co-ordinated project information is the 'common arrangement of work sections' (CAWS), which has superseded the traditional subdivision of work by trade sections. CAWS lists around 300 different classes of work according to the operatives who will do the work; indeed, the system was designed to assist the dissemination of information to sub-contractors. This allows bills of quantities to be arranged according to CAWS. Items coded on drawings, in schedules, and in bills of quantities can be annotated with reference back to the specification. Under the co-ordinated production information it is very clear that the specification is the central document in the information chain (Figure 2.1).

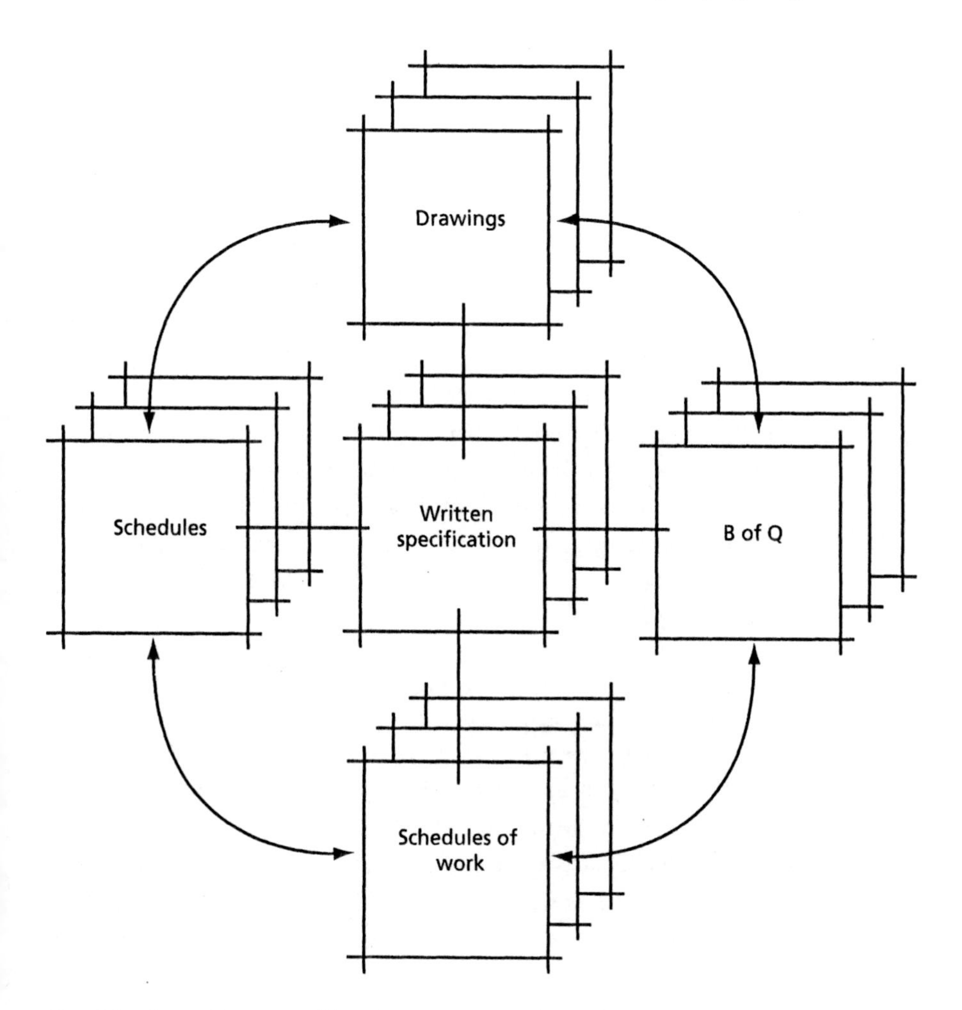

Fig. 2.1 The written specification: central to the information circuit

Drawings

Drawings are the most familiar medium and are regarded as one of the most effective ways of communicating information. Production drawings ('blueprints') are the main vehicle of communicating the physical layout of the design and the juxtaposition of components to those responsible for putting it all together on site. Referred to as contract information or production information, this set of drawings is usually complex and extensive. Not only does it take a great deal of time and skill to produce the drawings and co-ordinate them with those produced by other consultants; it is also a skill to read all the information contained and encoded in lines, figures and symbols. It is this set of drawings that the main contractor will use to cost the building work and (subject to any revisions prior to starting work) will be the set of drawings from which the building will be assembled. At its most basic, the contract drawings will comprise drawings produced by the architect, the structural engineer and the mechanical and electrical consultants. Other contributors to this set of drawings may include interior designers, landscape designers, specialist sub-contractors, highways consultant, etc.

Co-ordination of drawings with other consultants' information and the specification is an important consideration. A drawing system that aims to reduce repetition and overcome defects in less well co-ordinated systems is the elemental method. The elemental method is based on a four-category system, starting with the location drawings, focusing on the assembly drawings, then the component drawings and finally the schedules. Each drawing has a code and a number relating to the CI/SfB construction classification system. There are four codes, namely 'L' – for location, 'A' – for assembly, 'C' – for component, and 'S' – for schedule. This system allows specific reference to drawings and schedules to be easily incorporated in the specification, thus aiding co-ordination. Another aid to co-ordination is the use of consistent terminology, clear cross-referencing, and avoidance of repetition.

Specification notes on drawings

At this juncture, it is necessary to comment on the practice of writing notes on drawings. On very small projects and alteration works to existing buildings, it is common practice to write specification notes on the drawings, using a standard written specification to cover only the typical clauses common to most projects. Although widely used as a means of conveying information to the contractor, it is not good practice because the drawing very quickly becomes overloaded with information, repetition is largely unavoidable, and the majority of the notes are rarely descriptive enough to cover all the information required. There is a real danger that those reading the drawing on-site will rely entirely on the (incomplete) notes on the drawing and will not refer to the written specification, as they should. Apart from the obvious dangers of ineffective communication between designer and builder, this means that the drawing must be revised and re-issued every time there is a change to the specification, no matter how minor. It is considered best practice to keep notes on drawings to an absolute minimum and keep the written description of materials (size, colour, manufacturer, etc.) and workmanship firmly where they belong – in the written specification.

Written documents

Written documents have always taken precedent over drawings. Until relatively recently, it was common practice to award contracts on little more than a written description of what was required (indeed, this is still common in domestic repair and alteration work where the client directly employs an organisation to do work on their property, for example replacing the windows). The advantage of written documents, theoretically at least, is that people can understand them easier than they can drawings. Of course, this assumes that the document is well written and easy to read. Computers and word processing software have made the task of preparing and transmitting written material relatively easy; unfortunately, it is just as easy to proliferate errors.

Schedules

Schedules are a useful tool when describing locations in buildings where there is a repetition of information that would be too cumbersome to put on drawings. Particularly well suited to computer software spreadsheets, a schedule is a written document that lists the position of repetitive elements, such as structural columns, windows, doors, drainage inspection chambers, and room finishes. For example, rooms are given their individual code and listed on a finishes schedule that will relate room number and use with the required finish of the ceiling, walls and floor.

Schedules of work

It is common practice in repair and alteration works to use a schedule of works. This document describes a list of work items to be done. It is a list that the contractor can also use for costing the work. It is common practice to append the schedule of works to the specification, but it must not be confused with the specification (see below) or, for that matter, schedules (as described above).

Bills of quantities

The bills of quantities are derived from the drawings, schedules and specification. Usually composed by the quantity surveyor or cost consultant, their purpose is to present information in a format that is easy for the contractor's estimator to price. On small projects, it is unusual to produce bills of quantities because the information is usually concise, and the estimator can price the project from the drawings, schedules and specification. Bills of quantities are used on medium to large projects. Although computer software packages are available to generate the bills of quantities from the designer's information, it is still common practice for a third party to prepare them, and sometimes, this is the quantity surveyor. In doing so, the QS frequently finds discrepancies and omissions within the information provided, thus forming a useful (unpaid) cross-checking service for the design team. The contractor's estimator also has a duty to point out any deficiencies in the documentation to the design team.

Specifying quality

Drawings indicate the quantity of materials to be used and show their finished relationship to each other. It is the written specification that describes the quality of the workmanship, the materials to be used, and the manner in which they are to be assembled. Quality control should be foremost in the mind of the designer. It is a relatively straightforward task to set parameters for achieving quality through the written specification. This is vital because it is the contractor's agent, not the contract administrator, who has control on site.

Trying to define quality is a real challenge when it comes to construction, partly because of the complex nature of building activity and partly because of the number of parties who have a stake in achieving quality. Individuals and organisations concerned with achieving quality include government bodies (regulations, standards and codes), clients (individual and client bodies), manufacturers and suppliers (materials, products and systems), designers (architects, engineers, etc.), assemblers (craftsmen, tradesmen and construction managers), building users (often different from the client and changing over time), insurers and investors. Each group's perception of quality will vary depending on their particular position in the development 'team'. Not surprisingly, there is a great deal of confusion when people talk about achieving quality in construction, despite a growing body of literature addressed specifically at such issues. Usually, the word 'quality' is used in a subjective manner, rather than in an objective sense that can be tested and benchmarked. To confuse the issue further, provision of a quality service does not necessarily mean quality work; nor does a quality building necessarily have to be the product of a quality service. The two should, however, be inseparable (Maister, 1993).

Before looking at quality levels for materials and workmanship, it is necessary to agree on the quality level of the finished building. The projected life of the building and its use are primary considerations here and should be determined (as far as possible) at the briefing stage. So, too, should the maintenance strategy, life-cycle costing and disposal strategy for when the building has exceeded its design life. Obvious influences on the quality of the completed building are the client's budget and time frame, the composition of the design team, and the choice of contractor. More subtle influences concern the way in which the individuals party to a building project communicate the quality of the project information and the way in which the entire process is managed.

Quality of materials

Designers can define the quality of materials they require through their choice of proprietary products or through the use of performance parameters. In practice, the determination of quality is a rather complex issue because there are a range of characteristics that must be met, and it is usually necessary to place these in some order of importance for a particular project, as discussed in Chapter 5. Specifiers have to make up their own minds with regard to quality and may not necessarily take the same view as the contractor. The contractor is primarily concerned with getting the contract, thus submitting the cheapest price to get the job and then to maximise profits during the course of the contract, often through the substitution of products that generate a greater degree of profit. Quality is also dealt with, to a lesser or greater extent, through the use of the master specification combined with the use of approved and prohibited lists of products. The specifier's palette of favourite products could also be viewed as a mechanism for ensuring quality. These issues are taken up in greater detail below.

Quality of workmanship on site

No matter how good the detail design process and how comprehensive the resulting information, the quality of the finished building depends upon those doing the assembly and those doing the on-site supervision. Designers specify the position, quantity and quality of the building work – they do not tell the builder 'how' to construct it – this is the contractor's responsibility (hence the need for method statements). Choice of procurement route is an important consideration because it will set the contractual obligations of the designers and the contractors with regard to site supervision. With traditional forms of procurement, there are contractual obligations for the project administrator to inspect the work. With contractor-led procurement routes, such as design and construct, the designers may have very little legal say over the quality of construction since it is the contractor who is able to adjust specifications and substitute products without reference back to the designers. For the specifier, the decision to specify using prescriptive and/or performance methods cannot be separated from the type of contract used.

There are two principle users of the written specification on-site, the contractor's site agent and the clerk of works (and/or contract administrator). The written specification is not read frequently enough by site operatives; nor is it understood sufficiently, a situation exacerbated by the use of temporary operatives. Operatives will rely, for the most part, entirely on drawings and tend not to look at the written specification unless ordered to do so by the contractor's agent. Even if they do read the specification, it is unlikely that the standards referred to will be available on site. This may be habit and over-reliance on drawings, but it may well be grounded in a history of poorly written (and hence largely useless) written specifications. In an Australian study, it was found that bricklayers were not familiar with the Australian masonry code and that their work (perhaps not surprisingly) did not comply with it (Nawar and Zourtos, 1994) despite reference to it in the written specification. As soon as there is a problem or difference of opinion on-site, everyone reaches for the written specification to check whether or not the work complies with the standards expressed in the document. For those on-site, the challenge is to be familiar with, and have ready access to, relevant standards and codes of practice.

Prescriptive specifications

A prescriptive specification describes a product by its brand name, a proprietary product. For example, a facing brick would be specified as 'Ibstock, Red Rustic'. This gives the name of the manufacturer (Ibstock) and the name of the brick (Red Rustic) from the manufacturer's extensive range. It automatically gives the performance of the brick in terms of size, colour, texture, durability, water absorption, frost resistance, etc., as defined in the manufacturer's technical specification. This method of specification is usually quicker for the specifier than using the performance method and is favoured by designers for materials that will be visible when the building is completed. In the example above, the specifier has been precise and knows exactly what to expect from his selection in terms of appearance and quality.

Manufacturers spend a great deal of time and money in developing new products and/or improving products for which they usually hold patents. This does not stop their competitors from launching very similar products that are cheaper because they have incurred less research and development costs. Whether these alternative products are cheaper because they are of inferior quality is a point for debate, but one worth bearing in mind because there may be considerable pressure to change the specified product during the tendering and building phase of the contract. Furthermore, simply because a specifier has gone to a lot of trouble to carefully select and specify a proprietary product, it does not necessarily follow that the prescribed product will have been used on site. Specification substitution is a major cause for concern, for both specifiers and manufacturers.

One of the inherent problems with prescriptive specification is that of anti-competition. For public works, the practice is not to use proprietary specifications, the principle being that all relevant manufacturers must be able to compete for the work should they wish to do so. This is the case in the United States and Europe. Some large organisations follow similar principles by adding the words 'or equal approved' to allow some choice and competition between suppliers. Proprietary specifying is seen as being uncompetitive and not providing value for money; furthermore, the spectre of corruption is a difficult one to shake off. Performance specifying and compulsory competitive tendering are seen as a way around the problem. However, proprietary specifications are widely used and for very good reasons. For example, the recent trend towards the use of supply-chain management techniques necessitates a working arrangement with chosen suppliers. Although performance standards are used, this arrangement naturally leads to the specification of proprietary products from a particular manufacturer in the chain. Over time, the designers and manufacturers can work together to improve standards and reduce costs, but such arrangements make it difficult for others to compete.

Proprietary specification (no substitution)

In specifying proprietary products and not allowing substitution, the designer has made a choice and given the contractor precise instructions as to what to use. Responsibility for the specification rests with the design organisation and the implication of not allowing any substitutions is an implied guarantee that the product is fit for purpose and represents good value for money. Warranties, guarantees, and insurances should be sought from the manufacturer to transfer the implied liability from the specifier's office. The clause 'or equal approved' may be added (see below) to provide some latitude for change. Under traditional forms of contract, the contractor cannot make changes without the permission of the contract administrator. Given that a lot of time and effort will have gone into choosing a particular product in the first place, many specifiers are reluctant to change their specification without a very good reason, for example, a problem with delivery or an unforeseen technical difficulty on site. When changes are unavoidable, care should be taken to acquire and check the manufacturer's warranty before issuing the necessary instruction to the contractor in accordance with the contract. Because many changes are made under time pressures, this is not always done in practice but sought after the event; again, it is not good practice and should be resisted.

'or equal approved'

This clause is common in the majority of prescriptive specifications. By adding 'or equal approved', the contractor has some latitude in changing specified products as long as they are 'equal', and approval is sought from the contract administrator before the change is implemented. The design organisation remains responsible for the final choice of product and has a responsibility to check that any alternatives suggested by the contractor are fit for purpose and equal to that originally specified. Requests for changes must, under the contract, allow the contract administrator sufficient time to consider the proposed alternative. There is also a requirement for the contractor to provide the contract administrator with sufficient information (i.e. the relevant technical details and cost) from which a considered decision can be made. Too often, the contractor merely submits a list of products assuming (or hoping) that the design office has the relevant information. If products are unfamiliar to the office, then literature and samples will need to be sought, which takes time and can be arduous. Care should be taken to ensure that any cost savings are fully documented and passed onto the client (not the contractor). The use of 'or equal approved' can lead to arguments as to whether or not the product is 'equal' (in practice, some characteristics will be, others will not – hence the arguments). Adding 'or equal approved' is one way of dealing with the anti-competitive badge given to proprietary specifying, but we maintain that if a specifier has gone to a lot of trouble to select a particular product, to add such wording would invite substitution and potential problems. Many specifiers use the term 'or similar approved', which is not the same as 'or equal approved' and should be avoided.

'or equal'

Some designers specify proprietary products and then add the wording 'or equal'. This is an open invite to the contractor to use alternatives without asking for approval and is not considered to be good practice. It is used in an attempt to shift liability for product selection from the specifier to the contractor. If the specified product is used, then the specifier's office remains fully responsible, but if the contractor substitutes a product, then liability is transferred to the contractor. By using the words 'or equal' (and not using the word 'approved'), the contractor does not have to seek approval from the contract administrator and can substitute at their own risk. And substitute they will. Uncontrolled substitution for cheaper products is a certainty, and cost savings will not be passed onto the client because the substitutions will have been assumed at tender. The danger with using 'or equal' is that clients do not get what they think they are paying for: it is the contractor who profits. An additional concern is the quality and long-term durability of the building because substitutions will be made to suit the contractor, who may not be entirely clear as to the original design intent. In an American study of construction claims, litigation or arbitration related to specifications, nearly 25 per cent of the cases were related to the use 'of equal' clauses (Nielson and Nielson, 1981), which was double that of the next most common problem, ambiguous phrasing.

If the term 'or equal' is to be used (and we would recommend against it), then the specification should be written to require notice of substitution. This will allow substitutions to be tracked and provide information for an accurate design record of the completed building to be drawn up. There must also be a requirement for the contractor to provide a fitness for purpose warranty. In such situations (of doubt), a performance specification may be a better option because it transfers the actual choice of product to the contractor but at the same time sets defined parameters of quality.

Performance specifications

Unlike prescriptive specifications, performance specifications do not identify particular products by brand name; instead, a series of performance characteristics are listed (essentially a technical brief), which must be met. Performance-based specifications vary in their scope. They can be used to describe a complete project, one or more systems, or individual components. For example, clients may produce a performance specification for a design and construct project, engineers may produce a performance specification for the mechanical and electrical specification, and designers may specify such items as fire resistance and thermal insulation by the performance required (which is quantified). Continuing with the example of the brick, a performance specification would state the required size, colour, texture, durability, water absorption and frost resistance. Depending on the performance standard set, a range of different manufacturers' bricks may satisfy the required performance. This leaves the choice of product to the contractor and is popular in contractor-led procurement routes, such as design and construct, because it gives the contractor greater flexibility over product selection. Whether or not the contractor chooses Ibstock's Red Rustic (or whatever else the designer may have had in mind) will depend to a certain extent on how tightly the performance requirements have been written.

In passing the choice of product to the contractor, the contractor is given a design function and in doing so is expected to exercise reasonable care and skill in the same way as a designer. With performance specifications, it is not uncommon for the contractor to make last-minute changes to products in order to save money or meet programme deadlines. Care should be taken to record the final product choice, both as evidence in the event of a claim and for reference when alterations are made to the building. It is also important to check that the product selected complies with the performance specification required; sometimes, they do not, the contractor hoping that the specifier does not have time to check thoroughly.

Performance specifications were pioneered in the USA through their use in a school building programme in California in 1961 and started being used in the UK at much the same time (Cox, 1994) in public housing schemes and the provision of school buildings. The Property Services Agency (PSA) has been one of the leading advocates for performance specification in the UK, although it recognised that some elements were best specified using the prescriptive method. During this period, it was claimed that performance specifications were an effective way of improving efficiency and encouraging innovation: the reality was quite different, and the use of performance specifications declined rapidly. Performance specifications are best suited to large projects, while on small to medium-sized ventures it is common to use descriptive specifications for the majority of the work, with performance specifications for items such as the heating and ventilating system (HVAC). Over the past couple of years, there has been increased interest in performance specifying, and once again, the argument that it encourages innovation and reduces cost is being promoted, although there is little research to back such an argument. Interest in performance specifying has also come from those concerned with sustainability, inherently linked to the concept of life care (Wyatt and Emmitt, 2000).

Performance specifications are generally regarded as being more difficult and time-consuming to write than prescriptive specifications, but they are used for complete buildings (e.g. design and build contracts) and for sub-systems (especially building services and prefabricated building systems). One of the main challenges for the designer is deciding on the level of performance required: too narrow, and the tender is given little latitude; too wide, and the scope becomes too great to make sensible comparisons from the solutions presented by competing contractors. Care is needed to establish levels of performance that suit the project and the client. Performance specifications tend to be used by client organisations keen to leave the choice open (in the hope of getting the same performance cheaper than if a proprietary product was used). One argument put forward is that performance specifications are more effective in ensuring constructability and hence value for money on behalf of the client. There is little evidence to support this claim; indeed it could be argued that a good design team could ensure constructability and value for money through the use of prescriptive specifications. It would be misleading to suggest that one method is better than another, rather that different situations require different approaches.

Open and closed performance specifications

Performance specifications are sometimes referred to as 'open' or 'closed', depending upon the amount of latitude provided by the required performance criteria. An open (loose) performance specification would be written in such a way as to allow a great deal of freedom of choice for the contractor. A closed (tight) performance specification would be written in such a manner as to severely limit the choice of the contractor, sometimes to one or two possible products. Some specifications are so open as to render them worthless. However, some are so closed (usually because a manufacturer's performance specification has been copied from their technical literature by the specifier) that a descriptive approach would have been less time-consuming for all concerned. The designer's office is responsible for the performance specification because, if met, the solution will be fit for its purpose. The contractor remains responsible for ensuring the solution meets the performance criteria.

'Open' specifications

In addition to the use of prescriptive and/or performance specifications is the 'open' specification. These vary from the unintended (and possibly negligent) to their deliberate and considered use.

Open (silence)

Where there is silence in the documents, i.e. a particular item is not specified; it is referred to as an 'open' specification – a situation that usually arises because the specifier has forgotten to specify a particular item. For very minor items, most contracts allow for such situations, and the contractor assumes responsibility by choosing a particular product to suit. However, for major items missing in the specification, then the specifier's office is probably negligent for not specifying, and responsibility lies with the design office. Clearly, it is difficult to decide what represents a minor item and what represents a major item and caution is required. Careful checking procedures should limit the number of omissions in the documentation. An instruction will be necessary to rectify the silence, and there will be an additional cost to the contract that someone has to pay for.

Open (qualified silence)

Qualified silence is different to silence. An example of qualified silence would be a description such as 'use an approved undercoat', which, although considered, is another way of not specifying, some would argue a lazy way of specifying. Responsibility remains with the design organisation as long as the contractor submits details for approval, essentially, a way of delaying a decision and one that may slow down the contractor if approval takes some time to resolve. In the majority of cases, it would be better to specify a proprietary product or use the performance method.

Different approaches to specifying

As intimated above, there are a number of different approaches to specifying. We have found that the approach tends to be determined more by office policy and individual preference than by the professional background of specifier or office. Here, we look at the process from the perspective of four different specifiers drawn from our interviews with specifiers working in different design offices.

Four specifiers

The behaviour of four specifiers working in different design organisations helps to illustrate different approaches to specifying. All four approaches appear to work for the specifier interviewed and their design organisation.

1. The first specifier rarely used performance specifications, preferring to use the prescriptive method. Products were specified by proprietary brand name. Any requests from contractors to change specified products were vigorously resisted by this specifier and other specifiers in the design office, simply because it was their office policy to do so. Their argument was that 'the clients should get what they are paying for'. Prescriptive specifications were seen as essential in controlling the quality of the building because the materials used were dictated by the designer, not the contractor. This design organisation operated a very formalised list of approved products and a list of prohibited products that specifiers had to adhere to. This was part of their quality management system and helped to ensure that good-quality materials were always specified. The feeling in this office was that the designers must choose the products and then stick to them if quality was to be achieved. One of those interviewed summed it up with the comment 'why bother employing a designer if you are going to let the contractor choose the products, and hence determine the quality?'. This particular design office had a reputation for producing good-quality buildings that were cost-effective.
2. The second specifier used a mixture of prescriptive and performance specifications. She only specified by proprietary name when products were to be seen, i.e. when they formed part of the internal or external finishes of the building, such as facing bricks, roof tiles and internal finishes. Products hidden from view on completion, such as loadbearing blockwork, were specified by the performance method. Her particular design office did not have a standard policy on specification that all designers followed; instead, there was a wide variation in individual preference for specifying. Other designers within this organisation also varied their use of performance and prescriptive specifications to suit the project and their client. There

was no office dictat on specification writing; indeed, the whole office was very loosely managed. There was no office master specification; the universally adopted habit in the office was to roll project specifications from one project to the next. This office also had a good reputation for the quality of its buildings but was also well known for its inability to finish a project on time and frequently over-spent the budget. The specifier claimed to know that rolling specifications was not good practice, but that it worked for this particular office.

3 The third specifier, as a matter of office policy, specified entirely through the use of performance specifications. He rarely specified by brand name unless forced to do so, leaving the choice of product to the contractor and sub-contractors. Examples given for using prescriptive specifications were to suit project-specific requests from planning officers and client requirements. Because the performance specification was written for the majority of the elements specified by this designer, it was easy to draw from the master specification or previous project specifications as appropriate. Not only was this quicker for this individual to implement, but he was keen to keep building costs down for his clients. He believed that this method allowed the contractor to select the cheapest products that met the specification, and proprietary products would have been more expensive. This specifier had no concerns over quality, claiming that the secret to quality was selecting the right contractor for the job.

4 The fourth specifier worked in a large organisation that had set procedures for specifying. According to their 'standing orders', specifiers had to specify by proprietary name and should not use 'or equal' or 'equal approved'. However, the specifier claimed that it was standard procedure for specifiers to encourage the contractor to change products at the tender stage in an effort to save money. This made it look, on paper at least, as if they were saving the client money because of the difference between the initial cost estimate and the tender sum. The clause 'or equal' was used in the written specification, and the contractor was encouraged by the design office to take full advantage in order to reduce costs for the client and maximise profits. The priority for this designer (and his colleagues) was to keep the initial cost of the building as low as possible, his attitude was that one product was much the same as another, and thus substitution by the contractor was not an issue. He was aware of the fact that some of the products substituted may need to be replaced sooner than those initially specified, but since his client sold the buildings on completion, it was not his, or his client's, main concern. There was little documentary evidence to show what products the contractor had or had not changed; nor was there any real effort to pass the cost savings to the client. Rolling specifications from one project to the next was common practice. At the time of the interview, the office was having difficulties with the quality of their finished buildings and was faced with a number of litigations.

In addition to illustrating different approaches to specification, these examples also demonstrate differences in approach by different offices. Some specifiers hold very strong views on how to specify, as do some design organisations, whereas others are more relaxed in their habits and procedures. When conducting the interviews, there was a certain amount of surprise that someone (even a researcher) should be interested in the specification process. As one interviewee observed:

> Specifying is something you have to do very quickly, there is little time to think too long about the whys and wherefores – the only time we really give it much thought is when there is a problem.

None of the specifiers interviewed felt that the specification process was pivotal to good-quality buildings, maintaining that other tasks (such as detailing) were more important (a view consistent across samples of architects, technologists and surveyors). We beg to differ, especially given the amount of poor practice that we found in the course of the interviews.

There is a suggested relationship between specification practice and the resulting quality of the various offices' products. However, more research is needed to substantiate such a link.

External influences

There are a number of external factors that are under the control (to a lesser or greater extent) of persons/organisations located outside the specifier's office that will influence the specification process. This includes factors under the control of the manufacturer and others. The former are the cost of using the product, the availability for delivery to site on programme, associated services provided and the quality of the product information. The latter are other members of the temporary project 'team', such as the client's preferences, the influence of the town planner and limitations imposed by the main contractor.

External contributors

Architectural practice has a special relationship with one manufacturing industry, the building industry (Gutman, 1988). But this is not a simple homogeneous industry; it is made up of building-product manufacturers, professional consultants and contractors. Indeed, there is a clear distinction between the manufacture of the building products and their assembly. The manufacturers of building products will have an influence on the specification process, partly from the range of products that they produce but mainly through the communication of information about their products to the decision-maker, a process generally known as marketing.

Mackinder (1980: 133–146) listed a number of factors that influenced the selection of building products, which she called *external influences*, namely, the client, the quantity surveyor, the contractor, the government and the role of structural, mechanical and electrical consultants. More recently, the Barbour Index has investigated the role of different individuals in the specification process (Barbour Index, 1993), which was followed by further research into the influence of the contractor (Barbour Index, 1994), the client (Barbour Index, 1995) and the manufacturer (Barbour Index, 1996) on product decisions.

The client

The client will determine the overall quality of the building by both the amount of money and the amount of time available, thus defining the limits within which the architect has to work (sometimes exceeded). Mackinder's (1980:133) work found that certain clients do influence design decisions and hence product selection more directly, mainly through the use of a client's standard specifications, although four of the architects that she interviewed believed that the client should not be involved in the specification process. Research by the Barbour Index (1993, 1995) has indicated that the client has a major role to play in the selection of products, although the clients did report that they would rather not be involved if they could avoid it. In the 1993 report, 9 per cent of specification decisions were actually taken by, and 35 per cent influenced by, the client.

Consultants

At its most simple, the design team usually includes a structural engineer and a quantity surveyor in addition to the architect, and is influenced by the client's budget and complexity of the project. Generally speaking, the larger the project, the greater the number of different consultants who may, to lesser or greater extents, contribute to the individual specifier's decision-making process. For example, the structural engineer is responsible for advice on the structural system to be used, and will specify the structural items (Mackinder, 1980; Barbour Index, 1993).

Direct economic advantages or economic disadvantages to the client are different from those of the specifier, who may not always be aware of individual product costs. The quantity surveyor offers cost advice to both the client and the architect. Because of this, it may be that the specifier is not always aware of the cost of individual building products. Instead, it will be the quantity surveyor who has access to such information, both from information provided by product manufacturers and from experience from previous projects (indeed, it may be the quantity surveyor who will suggest alternative, cheaper, products to the specifier). Three quarters of Mackinder's (1980) sample of architects said that they left some items to be specified by the quantity surveyor, usually because they were short of time. However, the quantity surveyor was seen by the majority of the architects interviewed as a 'cost information file' and was asked to suggest alternative solutions when cost was very restricted. The Barbour Index (1993) found that the quantity surveyor was responsible for specification in only 2 per cent of cases and influenced the process in 10 per cent. This may suggest a change in practice, although the methods of data collection were different, which makes comparisons difficult and potentially misleading.

In addition to the formally constituted design team, the proposed scheme must have planning permission and building regulation approval. Legislative control may influence the selection of materials, especially materials that form the external fabric of the building. The architects interviewed by Mackinder (1980:141) said that the Building Regulations affected their selection of materials where fire prevention and thermal insulation was concerned. Architects also said that the planning officers exerted a 'very strong' influence on the external appearance of buildings and hence the choice of products in three-quarters of her sample. This was highlighted by the comment:

> Planners were also described as often being in the habit of stipulating proprietary products in order to achieve a particular effect known to them; examples quoted of this type were *Velux* rooflight windows and *Dorking Brick* (Mackinder, 1980:143).

Therefore, the town planner's contribution to the specification process would appear to be important where products are visible on the exterior of a building. There are clear dangers here as the planner has no responsibility for the building's performance.

The contractor

The main contractor as actual purchaser of the building products has a different relationship with the building product manufacturers than the specifier. Most contractors have accounts with one or two builders merchants, who themselves will stock a range of popular building products, and by placing their orders with a particular merchant, they can obtain some financial discount. Problems tend to arise when the specifier has specified proprietary products that are not stocked by the builder's normal merchant, in which case, the contractor may request a change of product to one that is familiar to him and is available from his regular merchant, but not necessarily familiar to the specifier. The issue of specification substitution is explored further in Chapter 6.

Constant variables

The external influences described above will exert different pressures at different stages in the design process. Furthermore, evidence would suggest that the other players in the network (such as town planners) may also have their own palette of favourite products, which, for various reasons, they will want to use on a project. This may be financially motivated in the case of the contractor and personally motivated in the case of the planner. These external influences could be described as a set of constant variables because they will exert pressures, to a lesser or greater extent, on every project. The extent to which these factors influence the specifier's decision-making process is explored and illustrated in the case studies reported later.

Information sources

The volume of information available to the building designer is vast. Each new project brings with it a new set of challenges and a fresh search for information to answer the questions raised during the design stages. This chapter provides an overview of the resources available, from information provided by regulatory and advisory bodies to that provided by the manufacturers. We also consider the information held within the specifier's office encoded in previous projects, office standards and master specifications.

Sources of technical information

Each new project brings with it a new set of challenges and a fresh search for information to answer specific design problems. Consider a specifier working in a design office. He or she will be engaged in a range of activities concerned with the design, detailing and production of a building project. In smaller offices, specifiers may be working on two or more projects concurrently, which may be at different stages. Product selection is carried out within this environment, and thus the capacity to access relevant information quickly is an important requirement. The individual designer cannot, and should not be expected to, be able to survey the whole body of literature available. Instead, an easily accessible, accurate, concise yet comprehensive, body of information is required. Some of the main sources for designers working in Britain are:

- Building Regulations and Codes;
- British (BS), European (EN) and International (ISO) standards;
- British Board of Agrément (BBA);
- *Building* Research Establishment (BRE) publications;
- manufacturers' technical literature;

- compendia of technical literature, e.g. *Specification*, *RIBA Product Selector*, *Barbour Index*, etc.;
- trade association publications;
- technical articles and guides in professional journals;
- previous projects worked on by the organisation;
- office standard details and master specifications;
- building information centres (UCIB).

This information has traditionally been held in the design office library as paper copies. More recently, many design offices have moved towards electronic information, held on CD-ROM or accessed on-line, usually by subscription to the provider. This makes access quicker, and, assuming the subscription has been kept up, the material should be up to date and reliable in content. There are a large number of interrelated sources of information that the specifier can draw on, explored below.

Testing and research reports

Independent research and test reports published by recognised building research organisations and papers contained in peer reviewed academic journals are the best source of information.

The British Board of Agrément

An Agrément Board was set up in Britain in 1966, modelled on the French Government's agrément system that already had an established track record. In 1982, the Agrément Board became the British Board of Agrément (BBA). This is an independent organisation that is principally concerned with the assessment and certification of building materials, products, systems and techniques. The status of agrément certificates is defined in the *Manual to the Building Regulations* 1995. Certificates guarantee compliance with the regulations where health and safety, conservation of energy and access for disabled people are concerned. Specifying a product or system that carries an agrément certificate will give assurance that the system or product, if used in accordance with the terms of the certificate, will meet the relevant requirements of the building regulations. Products assessed by the BBA are usually new to the market or are established products being used in a new, or innovative, way. For some products, such as thermal insulating materials, the Building Regulations require all new products to be certified by an independent organisation before they will be approved by a local authority. An agrément certificate, British Standards

Kitemark and or the CE Mark are therefore essential for companies hoping to sell their products.

Building Research Establishment (BRE)

One of the most highly respected research organisations is the Building Research Establishment. Originally set up as the Building Research Station and government-funded, this organisation was recently privatised. The BRE publishes a range of informative material on a wide range of technical issues relating to construction. These include the BRE Digests, BRE Defects Action Sheets, BRE Good Building Guides, BRE Information Papers and BRE Reports. It also holds a significant amount of information relating to all issues of building in its library. It has been, and continues to be, a comprehensive and reliable source of information for designers.

Manufacturers' own standards

Manufacturers set their own standards for production, delivery and after-sales service. The majority of manufacturers work to the current standards, simply because if their products do not comply with these, they would not be specified. Some manufacturers set higher standards than those within national and international standards, because they have the manufacturing expertise to do so. Combined with installation by their approved fitters, they can guarantee standards of performance and provide the necessary warranties and insurances that specifiers require.

Regulations

Prescriptive and performance specifications both rely heavily on reference to current regulations, codes and standards. Building regulations and building codes are first and foremost concerned with ensuring safe buildings and providing a healthy environment for building users. Regulations, in whatever form, offer designers a familiar set of controls to work with. They also offer a series of constraints based on experience, research and common sense – essentially a guide to best practice. National regulations, such as the Building Code of Australia or the Building Regulations *must* be complied with. Standards, such as International Standards (ISOs) and codes, such as the Code of Practice (CP) *should* be complied with. Busy designers may well view regulations, codes and standards as a burden because they are time-consuming to read and act upon. There is always the danger that a standard is referred to in a specification without the specifier necessarily reading the standard and/or fully understanding its subtle complexities. Nevertheless, one of the many skills of the building designer is his or her ability to keep up to date with current legislation, applying it in a creative and cost-effective manner to realise the design intent.

Research by Mackinder and Marvin (1982) found that building designers tended to refer to written documentation only when they had to, preferring to rely on rules of thumb and their experience until such time that they were forced to search for information. Clearly, there comes a point when someone with technical knowledge will have to make the design work, including compliance with legislation and best practice. In some design offices, this is carried out by the same person who crafted the conceptual design; in others, the task of detailing and ensuring the design complies with prevailing legislation will be carried out by more technically orientated individuals, technologists and specification writers.

Regulatory frameworks

Regulatory frameworks vary between countries, and regional variations are not uncommon in many countries. For example, in Australia, there are state and regional variations to the Building Code of Australia to accommodate specific regional conditions. Globally, the trend has been to move towards a performance approach in preference to a prescriptive one. Prescriptive regulations show or describe the construction required to achieve conformity. Performance-based regulations stipulate a level that must be met (or bettered) but do not specifically indicate how this is to be done. Another difference between prescriptive and performance regulations is that performance-based regulations focus on methods of conformity that consider buildings as whole systems, rather than elements in isolation. Thus, trade-offs between component parts, for example in achieving the necessary minimum thermal insulation values, are allowed to achieve the given objectives. Theoretically, the performance approach allows designers greater freedom of expression in the way in which conformity is achieved, although in practice, there is still a heavy reliance on standard details (usually copied from the regulations) to save time.[1]

In the UK, the first national Building Regulations were introduced in 1966, which brought all of England and Wales under a common legislation for the first time, replacing the varying local by-laws. The Building Act of 1984 led to redesigned regulations

1. Another approach is an objective-based one, as adopted in the National Building Code of Canada (NBCC), which should allow a greater opportunity to address matters of durability (Chown, 1999). An objective-based approach seeks to combine prescriptive, performance and functional requirements.

and a set of *Approved Documents* in 1985 that described construction that met the requirements of the Building Act. This represented a major shift from a prescriptive approach to a performance one. The *Approved Documents* are intended to provide guidance for some of the common forms of construction whilst encouraging alternative ways of demonstrating compliance under the 'deemed to satisfy' standards. Designers and builders now have a choice: they can accept the suggested method in full, in part, or not at all if they can demonstrate an alternative method of compliance. In reality, many designers and builders find it quicker, easier and more convenient to work to the solutions suggested and illustrated in the *Approved Documents*; alternatives are more time-consuming and may well be rejected, leading to delays. Since their re-design in 1985, there have been a number of revisions and additions to the *Approved Documents*, with a major revision in 1991, and ongoing updates. In their current form, the *Approved Documents* form a useful *aid memoire* and detailed design guide.

Standards and Codes of Practice

Standards and Codes provide guidance to designers based on best practice and research findings. They are an effective way of bringing research and development to practitioners, thus aiding the adoption of new technology and raising standards of quality. Standards and building codes are prepared by committees of specialists drawn from government, academia, manufacturing, professional practice, and user groups. The results are documents arrived at by consensus, and, as such, they may not meet (or indeed be seen as relevant to) the needs of all. National research organisations such as the Building Research Establishment in the UK and international research bodies such as the International Council for Research and Innovation in Building and Construction (CIB) are actively involved in the development of national and international standards through representation on various development committees. For example, the CIB have been active in the development of standards for sustainable construction through the International Standards Organisation (ISO).

Standards and Codes of Practice have two functions. On one level, they provide the designer with advice and guidance, and on another, they provide the specifier with a certain amount of security since they represent best practice. Working to both relevant and current standards, the designer will be safe in the knowledge that he or she is applying the most current knowledge. This reduces risk for the design firm, ensures the safety of those doing the construction, and protects the interests of the client. However, it should be remembered that many standards are developed in the light of failure, and problems may, unfortunately, still occur. Structural codes are re-assessed and usually revised in the event of a structural collapse. Designers working at the cutting edge of technology are likely to be ahead of the relevant standards since they do take a long time to develop and/or revise. Dangers can also arise when designers are, without realising it, working outside the limits implicit in the drafting of the code or standard.

New standards and codes

The development of new but similar products by different manufacturers results in a wide range of properties that can be confusing to the specifier. Nationally or even internationally applied Standards address this problem. The BSI was the first national standards body, but now there are more than 100 similar organisations.[2] Formed in 1901 (as the Engineering Standards Committee), its first British Standard concerned with building was published in 1903, standardising the size of rolled steel (Yeomans, 1997). Licences are now given for products to carry a 'Kitemark', which certifies that the product complies with the relevant BS. The first of these was issued in 1926, and British Standards have become an essential tool for building designers.

In 1942, the British Standards for Codes of Practice (CPs) for design were introduced to ensure a degree of uniformity. By designing in accordance with the Codes of Practice, designers were 'deemed to satisfy' the legal requirements of the time, and many became standard works of reference (Yeomans, 1997). Codes of Practice are based on a combination of practical experience and scientific investigation and form an essential part of quality assurance schemes. Since the early 1990s, Eurocodes have been published that establish standards for the design of structures across the European Union. Examples include Eurocode 5, *Design of timber structures* (1994) and Eurocode 8, *Design provisions for earthquake resistance of structures* (1996).

With increased attention on the international market has come a focus on International Standards. The International Organisation for Standardisation (ISO) was founded in 1947 with the aim of harmonising standards internationally. As with national standards, such as ASs (Australian), BSs (British) and DINs (German), the ISO series serve as guidance to designers and specifiers and do not necessarily have to be complied with. National standards are being replaced with European Standards (ENs) and ISOs, where appropriate; for example, the ISO series on quality, BS EN ISO 9000–9004 has superseded BS 5750. Likewise, the Australian and New Zealand codes are being replaced with ISOs.[3]

2. These belong to the International Organisation for Standardisation (ISO) and the International Electrotechnical Committee (IEC).

3. Three ISO series that are closely linked and important to architects offices are the 9000, 14000 and 15000 series on quality systems and management, environmental management, and design life, respectively; combined, they provide comprehensive guidance on important areas that designers and specifiers cannot afford to ignore.

Although standards and codes are in a constant state of development, the conservative behaviour of designers may result in their failing to keep up with developments in the latter. Keeping abreast of changes in standards is a relatively simple matter compared with keeping up with design codes. Learning a design method represents a considerable investment on the part of the designer. The result is often considerable resistance to the introduction of a new code of practice because this will mean that designers will have to learn new routines. Figures that the designers carry in their heads will no longer be relevant, and even the whole philosophy of design may have been changed. This is what happened when structural codes changed from a reliance of allowable stresses to load factor methods. Thus, where old codes have not been withdrawn, some designers persist in their use. Some structural engineers are still using BS 449, *The Use of Structural Steel in Building*, even though a new code has been available for many years. The same is true of CP3 Chapter 5, the code on wind loading that has been replaced by the much more complex BS6399 Part 2, and so still remains in use because of its relative simplicity.

Trade associations

The trade associations can usefully be divided into organisations representing general materials, such as the Timber Research and Development Association (TRADA) or the British Cement Association (BCA), those representing a type of product, such as the Brick Development Association (BDA), and those representing groups of specialist contractors such as the National Federation of Roofing Contractors. The first two types will produce information that is useful at the design stage. For example, TRADA produce Wood Information sheets that provide information on general issues such as the design of laminated timber or on the design of timber-based construction such as fire-resisting walls or separating floors. Some even publish standard specifications. The intention here is to help the designer at this important stage in the process improving the prospects that the material whose interests the association represents will be incorporated into the design. The designer's concern at that stage will be the performance of the component or assembly, and so it should be this information that will be found in the trade-association literature.

A quite different function of trade associations is to provide an assurance of construction quality. This might be done through a wide range of quality assurance schemes. The simplest is perhaps the regular testing of manufacturers' products, the advantage to the specifier being that the trade association offers a measure of independence and hence reliability. Examples of this are the TRADA schemes for trussed rafter roofs and fire doors where products assured within the schemes carry identifying marks for simple checking on site. Associations of specialist contractors might limit their membership to those who comply with certain standards set by the association, for example standards of training for their operatives. The assurance being offered here is that the work will be carried out in a satisfactory manner; something that may be difficult to check on site once the work has been completed or that may even require specialist knowledge to assess. Of course, the judgement that must be made by the specifier is whether or not the claims made by such associations are delivered in practice. Some years ago, the literature of one association implied that it employed a particular academic as a consultant. On enquiries, the person named denied any such link. In spite of this, such associations do perform an important quality-control function. In the UK, the majority of house-builders belong to the National House-building Council (NHBC) that sets its own building standards (some of which are more stringent than the Building Regulations) and training standards for members. In recent years, the organisation has made considerable progress in improving the quality of housing built by their members, for which new owners receive a guarantee.

The larger trade associations are usually represented on British Standards committees and will have an input to changes to the Building Regulations. They also have an input to the development of the Eurocodes.

Property is a major asset, and it should come as little surprise that the larger insurance companies exercise an influence over building standards. In addition to the NHBC's insurance scheme, many large building projects are vetted by insurance companies at the design stage to assess the amount of risk against their own guidelines for security and fire protection. Other organisations work to design guides such as the Housing Association Property Mutual *HAPM Component Life Manual* (HAPM, 1991), which gives extensive information and benchmarking for component service lives of materials and some M & E components. In France, the requirement for buildings to be insured means that insurance companies require that designs be checked effectively taking on the role performed by local authorities in the UK.

Manufacturers' information

Collectively, building product manufacturers produce a variety of information for a variety of uses, from promotional literature to technical literature. In many respects, the pattern books of the late 19th century have been replaced by manufacturers' own technical literature that often contain full working details showing the way in which their products can be used. It can even include typical specification clauses that can be easily used by designers. At the other end of the scale is promotional literature that is simply designed to raise awareness of the company and its products to the specifier. Rarely does it provide enough information to allow the specifier to actually specify the product; rather, the intention is that the specifier should make contact with the manufacturer to ask for further information. The manufacturer might then choose to send technical literature, deal with enquiries by telephone and/or send a technical representative to the specifier's office to assist with the specification. While the specific function of different kinds of manufacturers' information might be different, all raise the awareness of the specifier to the products described, and therefore, the general term 'trade literature' is used here for all.

Manufacturers have also tempted designers to use their new products through advertising in the professional journals. The first and subsequent editions of journals such as *The Builder* (1843) and *The Architects' Journal* (1895) have carried advertisements from a wide range of manufacturers selling an ever-wider range of products and services. Advertisements are important for the Journals, since the revenue generated by them helps to finance their production and distribution. Some of the 'product journals' that exclusively contain advertisements are distributed free of charge to specifiers' offices because the entire cost is borne by advertisers. Specifiers need to exercise a degree of caution in using technical features in journals because many are little more than a re-work of press releases of manufacturers, usually evident from the lack of any critical discussion of the components being described.

Trade literature is designed with the principle objective of helping manufacturers to increase their sales, and the consistency and quality are as varied as the products on offer. Building materials and products manufacturers are in business to sell their products to specifiers, and to do so, they must make them aware of their product range. From the earliest days of mass production, manufacturers have advertised their products to potential specifiers, the designers, builders and tradesmen through trade literature and advertisements in the specialist journals. Product catalogues were, and still are, an effective means of selling products because they are convenient for specifiers to use and order from, through the act of specifying them in the contract documentation. The more recent provision of typical details and specifications on disk, or more recently from the internet, which can be imported into the construction details, has been developed in tandem with the general adoption of computer-aided design in architect's offices. In many respects, information available through the internet is not significantly different from that offered in printed form; the difference is the ease with which it can be transferred to contract documents.

By making it easy and quick for the designer to import their standard details and standard specifications into project information, they can find their products specified without even being contacted by the designer. This 'free' information has been carefully designed so that by selecting a particular manufacturer's detail, the designer is also confirming his or her choice of product. The situation is a little more complex where performance-related specifications are being used, but manufacturers do provide performance specifications for their products that are so written as to confine the choice to their product only. Needless to say, in the longer term, manufacturers hope that the designer, and hence the design office, will adopt their particular detail and their product as a standard detail. Many manufacturers will also employ technical staff who will provide bespoke details for a particular project. A feature of manufacturers' own details is that their literature often contains errors. Manufacturers are concerned with the promotion of their own product range so that other components may not necessarily be represented correctly. As with all 'typical' details, caution should be exercised when working such details into the overall detail design drawings, because, once included, they tend to remain.

In common with standard details, the supply of a product specification is an effective method in helping the designer to adopt a particular product over that of a rival manufacturer. This is true of proprietary specifications and performance specifications, the latter being written in such a fashion as to make the selection of a rival manufacturer's product impossible. For example, a performance specification for a particular product may specify a manufacturing tolerance that rivals cannot achieve.

Product compendia

Product catalogues such as the *Barbour Index* in the UK and *Sweets Catalog Files* in the USA provide a convenient and familiar point of reference for busy specifiers. Along with other compendia such as the *ASC Files*, *Barbour Compendium* and *RIBA Product Selector* they are a compilation of individual manufacturer's building products and are published annually. These list products under general subject headings but do not cover all manufacturers, only those who pay to be in the directory. The compendia, available in both paper and electronic format, are not designed to offer advice on product selection, nor do they provide a comparative assessment of similar products; they merely list building products and provide generic descriptions of materials. There is no comparative advice on cost or performance. Thus, the specifier cannot refer to a publication that provides comparative product assessment (unlike, for example, the potential car purchaser, who can refer to specialist journals that provide comparative information about cost, performance and value for money). They do, however, provide a useful source of information, and the interactive web-based compendia allow the specifier rapid access to manufacturers' own web sites and additional technical information.

Specification, which was first published at the end of the 19th century, set out to provide guidance on specification. Set out under different trades, it described the processes of construction and provided specification clauses that architects could use.

Trade representatives

Representatives form an important link between manufacturers and potential specifiers of their products, but they are an expensive resource, and not all manufacturers employ them. They have a dual role, employed both to raise the awareness of the specifier to their employers' products (a marketing role) and to provide information and help to the specifier with the aim of getting the specification (a technical and sales role). For the purposes of this book, the term 'trade representative' is used to cover both sales representatives and technical representatives.

While trade literature tends to describe and illustrate typical solutions, in many circumstances, the specifier is faced with anything but typical situations. Thus, further information has to be requested from the manufacturer, and details may need to be discussed over the telephone or face to face with the manufacturer's trade representative. It is this situation that defines the specifier's requirements of a trade representative. However, like designers, the quality of trade representatives varies, from those with excellent technical knowledge of building and their organisation's product range to those who have limited experience of building but are good at selling. Naturally, the former are very useful to designers, but the latter are regarded as a waste of time because they are unable to answer the technical questions asked of them (furthermore, many designers do not like being sold to).

Of course, the quality of the trade representative may be determined by the policy of their employers. Little research has been done on this, and the information would doubtless be considered commercially sensitive, but some manufacturers appear to concentrate on the selling function, while others attempt to provide good technical support. The common attitude within the design community to trade representatives is either unknown to product manufacturers or ignored. The former is possibly because their advertising is designed by advertising agencies more familiar with other markets and unaware of the culture of architects' offices. This culture and the mechanisms by which offices handle trade representatives are dealt with in Chapter 8, but it is appropriate to summarise some of the findings of the research into this. The common view is that representatives are regarded as a waste of time by offices and, if they must be seen, are given the cheapest person to talk to – a junior member of staff who will have little influence on specifying policies.

Specifiers require technical knowledge and technical information to be provided quickly and accurately, usually for a very specific purpose. This means that trade representatives who have empathy with the designer's concerns and can answer their queries quickly will be influential in helping the specifier to choose their products over those of a rival manufacturer. Another strategy employed by manufacturing organisations is to provide a technical help line to answer technical queries. Sometimes, these are provided instead of trade representatives, sometimes in addition to them. Quick and accurate responses to specific technical questions will be expected by specifiers, and the speed and quality of the response may influence their decision to use or reject a particular product. In view of the fact that providing trade representatives is an expense for manufacturers, one would imagine that they would want them to be as cost-effective as possible. The clear lesson from the work that the authors have carried out is that it is both technical literature and technical advice that designers and specifiers require, the former at the design stage so that they do not need to waste time with technical enquiries, the latter at the specifying stage and produced in response to demand.

For many specifiers, the service provided by the manufacturing company and/or supplier is equally as important as the characteristics of the product. Help with detailing difficult junctions and writing the specification will be welcomed by busy designers with tight deadlines. Technical helplines and the prompt visit of a trade representative to assist and provide product-specific knowledge are important services that can give manufacturers competitive advantage over their immediate rivals. Typical services provided by manufacturers may include:

- guaranteed response to technical queries (within 24 hours);
- bespoke design service and provision of free drawings, details, specifications and schedules;
- provision of CAD files;
- structural calculations for submission to building control;
- on-site technical support;
- product-specific guarantees and warranties;
- access to accredited installers.

Efforts to develop a working relationship between manufacturer, designer and contractor are a small investment for all parties to ensure a relatively trouble free 'partnership'. These help to ensure future specifications for the manufacturer.

It would be misleading to give the impression that designers develop their detail designs and specifications in isolation. Successful design relies on co-operation between manufacturer and specifier. Manufacturers have a vital role to play in helping the designer to detail particular aspects of buildings, especially in circumstances where the detailing may be unfamiliar to the designer or to the design office. On large projects and projects with unusual details, many manufacturers will offer to provide the technical drawings and written specification clauses for the designers; for example, cladding companies will provide a complete package. This saves the design team a lot of production work, shifting their emphasis to co-ordination and checking information from other sources.

Touching the product

A service that manufacturers may provide, usually through their representatives, is to supply samples of their products. Of course, by the time that the specifier is requesting such a sample, a preliminary decision to use that particular product will already have been made. At this stage, the function of the sample is to confirm the choice rather than to select among alternatives; something that occurs at an earlier stage in the design. What the designer needs for this is a collection of samples from a range of competing manufacturers. A resource that has been available to specifiers in large cities has been the building information centres that provide a shop window for a variety of manufacturers to exhibit samples of their products and display their trade literature. They have always been popular with specifiers because they provide an opportunity to touch, assess and compare products without any pressure from the manufacturer's sales representative. However, the range of products on display is both limited by the size of the showroom and restricted to those manufacturers prepared to pay for the space. Thus, the products on display tend to be those of the larger manufacturers with larger marketing budgets rather than their smaller, less affluent, competitors.

Naturally, the ability to look at and compare different materials is particularly important for those products that affect the finished appearance of the building. The brick manufacturers, Ibstock, established a series of design centres in major cities throughout the UK as a means of bringing their bricks to a larger audience (Cassell, 1990). Other commercial concerns also operate showrooms for particular product types. For facing brickwork, there are a number of brick showrooms around the country. Known as 'brick factors', they act as a middleman, stocking a wide variety of bricks from different manufacturers and aiming to encourage the designer's specification.

Another source of information are the centres aimed at promoting environmentally friendly building products and systems. A well-known example is the VIBA exhibition at s'Hertogenbosch in the Netherlands, which displays materials and arranges educational events to raise the awareness of specifiers to alternative ways of detailing and specifying buildings with the aim of minimising the environmental impact on our planet.

Builders' merchants are another source of information about products stocked and available in a particular location. Although these depots may not be the designer's first port of call for information, for those keen to specify products readily available and clearly priced, they represent a good source. Again, the products are on display, although in less glamorous surroundings than the building information centres. For specifiers keen to use local suppliers and locally sourced materials to reduce the haulage (and associated pollution), the local builders' merchants provide a good source of information.

Clients' specifications

Some clients, especially those who have a large portfolio of buildings, often develop their own requirements, expressed as a 'client specification'. Client specifications vary in both their complexity and their use of performance and/or prescriptive methods. At its simplest, the client specification will be a list of performance criteria, perhaps supplemented with a list of materials that are not to be used. For example, a client wanting a new warehouse may simply list the floor area and volume required, together with specific requirements for loading bays and percentage of office accommodation to warehousing space. At the opposite end of the scale are client specifications that are extremely detailed. Such lists vary in complexity, depending upon the building type and the experience of the client organisation. For example, for buildings where hygiene is an overriding consideration, for example in food preparation, there will be some specific requirements that must be met. Such specifications represent a source of expert knowledge developed by the client organisation over time and revised to suit changing circumstances and improvements to their standard requirements. As such, they provide an excellent briefing document and detailed design guide from which to work. Where design organisations carry out repeat projects for such clients, it is standard practice to develop a bespoke master specification for that particular client.

Lists of products that should, or must, be used and possibly a list of prohibited products also develop from the client's experience, both good and bad. Such lists can be extensive. Although many design organisations take client specifications as a definitive list, some designers question them from time to time, especially in situations where there is discrepancy between what a client wants and what other organisations/control agencies may require to ensure conformity. For example, repair and maintenance work to a listed building may require the use of lead-based paint when the client specification clearly states that such paint should not be used. In the majority of cases, differences should be resolved quickly and incorporated into the project-specific specification.

Office standards and masters

Although specifiers are bound by professional ethics that prevent them from accepting any financial inducement for selecting a particular product, they are unique as consumers because they are selecting products that are themselves products of a design process, marketed through carefully designed advertising and technical literature. Therefore, there is the possibility that specifiers are more likely to select products that they can empathise with, i.e. products perceived as being most sympathetic to their design values (e.g. Grant and Fox, 1992). They may be more receptive to advertising literature that accords with their own design philosophy in terms of style, composition and colour, etc. Thus, competing products may not necessarily be analysed objectively, but subjectively. We should point out that there is a difference between the goals of designers and those of contractors. Contractors will be more interested in availability, buildability and profit margins – an important point to bear in mind when using performance specifications.

The palette of products

The specifier will have a repertoire of well-rehearsed solutions that work, based on experience acquired through education and practice and, it would appear, reinforced by a palette of favourite products, illustrated by the case studies below (see Chapters 9 and 10). There are a number of advantages and disadvantages to the use of personal information within the design office:

1 The advantages are that:
   - it saves time looking for information;
   - the product is familiar so that uncertainty is reduced;
   - details and specification clauses can be imported from a previous job.

2 The disadvantages are that:
   - it reduces the likelihood of specifiers looking for alternatives and so might hinder innovation;
   - there is a greater chance of superseded information being used;
   - there is a greater chance of error through the use of rolling specifications.

Many specifiers have their own favourite products and sources of information based on their individual experience. This is represented by a 'palette of favourite products', from which they always choose unless forced to do otherwise (Mackinder, 1980; Emmitt, 1997). Use of products from this personal collection reduces the amount of time spent searching for products to suit a particular situation, and because they are known to perform well (or more to the point known not to fail), their selection poses little risk to the specifier. This palette of product information is essentially a knowledge base that is used to aid the specification process. It may be maintained as a file of paper literature or in a digital file that can be quickly imported into drawings and specifications. This is an important information source and is investigated further in later chapters.

Quality management systems prohibit the use of individual files of product information because of the danger of it being out of date. However, research by the authors found that specifiers went to great lengths to maintain their own file of information. With the increased use of computers, the palette of favourite products can be easily stored in an electronic file. Quality managers need to ensure regular audits to either eradicate the use of personal sources or ensure that the information contained in personalised files is current and in accordance with office standards: the former is easier to manage than the latter.

Similarly, the majority of design offices build up some experience of successful and unsuccessful products and details over time. Wade (1977:158) has commented on the reliance of known sources:

> Each office develops its own library of product information and begins to develop preferences for the use of some products and not for others – as a result of good and bad experiences of those products.

This experience may be disseminated through internal memoranda or office standards leading to the development of an office palette of favourite products, possibly incorporated into standard details and the master specification. Such practice reinforces established patterns of behaviour and discourages independent thought or action, confining specification to products from the list of approved products and preventing the use of 'blacklisted' products, essentially forming a further barrier to the use of innovative products.

Standards and masters

In the majority of design offices, typical details and specifications are customised to suit the organisation and hence become office 'standards' or 'masters' that are used to save time and ensure a degree of consistency. These standards are based on good practice (as viewed by the design office) and the collective experience of the office. Some details and specification clauses will be unique to a particular design office; others will be amalgamations of typical details, typical specifications and manufacturers' details and specifications. The use of office standards by design organisations is common practice; sometimes criticised for stifling innovative design, they are used to save time and reduce the risk of failure. Not only do they save time in generating the same drawing or specification over and over again, but they also encourage good practice. As such, standards of this kind form an essential part of well-designed quality-management systems. They form part of the organisation's collective knowledge and can be used by less experienced staff, as long as the process is monitored and checked by a more experienced member of the office. This information can be broken down into two categories: those products that are approved and those that are not.

Approved and prohibited products and materials

Most design offices maintain a list of approved products. Products that are known to perform, through both care in their selection and subsequent experience in their use, are those included in the approved list. Such lists help the organisation to ensure quality and compliance with particular standards and regulations. As noted above, some clients insist that the design team use certain products or suppliers because they have good experience of using them.

Because of previous bad experience, some products and or manufacturers may be included on a prohibited list. Inclusion on the list may be because the product failed in use, was found to be of poor quality when delivered to site, or was difficult to use on site, or simply because the service provided by the manufacturer was found to be unacceptable. Poor service by the manufacturer may include a failure to respond to technical queries on site or failure to deliver products to site as scheduled.

The 'master' specification

Graphic representation of architectural details is used to control the form and appearance of the building project. Written specifications are used to control the quality of the materials used and the quality of the workmanship. It is the specification that controls quality, and the development of office standard specifications is equally as important as the standard details. These office standards are frequently linked to national and international standard specifications. In the USA, such a system has been developed by the Construction Specifications Institute (CSI). As with standard details, standard specifications are used by organisations to save time, merely adjusting the master document to be project specific. Given that many design offices tend to specialise to a certain extent, e.g. housing, offices, hospitals, etc., the development and maintenance of a carefully written master specification make sense because of the major savings in time and the reduction in risk. The master document needs to be maintained on a regular basis to retain currency and project-specific text added to suit particular circumstances for each job. This means that someone in the office must be responsible for the upkeep of the master specification, responsible for its regular review and ensuring that all alterations are checked and recorded in accordance with the quality-management system. In small offices, an individual will do this in addition to their other work; in larger offices, the task will be carried out by someone with responsibility for technical matters.

Selection criteria

Most books dedicated to specification writing fail to address the crucial decision-making stage during which individuals actually select certain products or choose certain parameters. This chapter investigates the various, and often conflicting, selection criteria that may be employed by specifiers to achieve their objectives.

Detail design decision-making

> the selection of mutually compatible components to solve some specific defined standard problem or some specific unique problem within a larger consistent design context (Wade, 1977:281).

At first glance, the selection of materials and products to meet a specific purpose would appear to be a relatively straightforward activity. However, on closer inspection, both the issues to be considered and the actual process are complex. One of the central problems facing designers is that of determining priorities. Resources are not infinite, so to achieve the given objectives within the constraints of time, finance and expertise, the number of variables considered has to be limited, i.e. designers must determine their priorities for each design project. Furthermore, the selection of one product over another will affect the selection of other components. No architectural component can be considered in isolation: the designer must constantly appraise and re-appraise the product in relation to the building assemblage as a whole. Detailing a building is a process of continual evolution grounded in what the designer (and the design office) believes to be best practice. Detailing is based on the sharing of information and knowledge pulled from a wide variety of sources and influenced by many different contributors. It is also about the choice of the correct solution for a particular set of circumstances at a particular time, considering the benefits for clients, builders and users within a framework of limited resources and creative endeavour.

Underlying all issues concerned with design, manufacture and assembly is essentially the ability to make decisions in the available time. Design is a particular type of decision-making activity, and one about which there has been much research and debate, although whether or not there is yet an adequate 'natural history' of the design process (Yeomans, 1982) is debatable. Heath (1984) and Rowe (1987) provide useful, and critical, insights into design thinking. A slightly different perspective is that of design as an information processing activity (Akin, 1986). This, and associated work, is based on the premise that designers, managers and hence organisations can be understood by observing their decision-making behaviour and then designing and implementing an information-processing model (Simon, 1969; Newell and Simon, 1972). Heath (1984) concluded that the appropriate method is determined by the social nature of the task, with different methods employed to suit a particular situation. Through the use of models, designers can select from a range of tactics to shorten the time required to produce the design and to reduce the potential for error. This observation is relevant to the conceptual design stage and the detail stage. Research into failures of detail design suggests that they arise because of an inability to conceive the building and its constituent parts as a dynamic system: not so much a case of ignorance, more an inability to apply the appropriate method (Heath, 1984).

Decision-making

Much of the literature dedicated to design decision-making is centred on the actions of individuals, with less emphasis placed on the collective efforts of individual organisations and the building 'team'. Clearly, it is easier to observe the behaviour of individuals, especially in controlled experiments, rather than the group activity of a design office. Therefore, this bias in the data available is to be expected. In practice, however, individuals are constrained and influenced by the behaviour of the group to which they belong and by other groups party to the design process.

When people make decisions, they tend to follow rules and/or procedures that they see as appropriate to the situation (March, 1994). This is particularly so of professionals who are expected to act in a certain manner according to their particular professional background. Not only do designers have to satisfy their client, but they also have to satisfy different building users and, as in many other professions, will feel a need for peer approval. Building designers are expected to act in a logical manner when selecting building products and materials, assessing all the options, against a background of legislation, before making a choice. Research suggests that this may not be the case. Although much of the literature on decision-making makes assumptions based on rationality, the validity of this assumption is thrown into question by studies carried out by behavioural scientists. Observational studies of decision-making behaviour suggest that individuals are not aware of all the alternatives, do not consider all of the consequences, and do not evoke all their preferences at the same time. Rather, they consider only a few alternatives and look at them sequentially, often ignoring available information (March, 1994). Decision-makers are constrained by incomplete information and their own cognitive limitations; thus, although decision makers may set out to make rational decisions, in reality, they make decisions based on limited rationality: they search for a solution that is 'good enough', not the 'best possible' solution.

The factors that place constraints on human decision making are:

1. Attention span. It is impossible to deal with everything at once. Too many messages, too many things to think about. Thus, we tend to limit our attention to one task at a time, ignoring messages that are irrelevant to that particular task, engaging our selective exposure. Our attention span is also limited by time.
2. Memory. Both individuals and organisations have limited memories. Memories are not always accurate; we tend to remember acts as we like to see them, rather than as they actually happened. Organisations and individuals are limited by their ability to retrieve information that has been stored. Records are not kept, are inaccurate or are lost so that lessons learned from previous experience are not reliably retrieved. Moreover, knowledge stored in one part of an organisation cannot readily be used by another part of that organisation.
3. Comprehension. Despite having all the facts to hand, the relevance of information may not be fully understood. There can be a failure to connect different parts of information. Furthermore, individuals have different levels of comprehension, making it difficult to foresee how each will respond to the information they have. For example, the architect, manufacturer and contractor may understand the same piece of information differently simply because of their different backgrounds.
4. Communication problems. Specialisation, fragmentation and differentiation of labour encourage barriers and present difficulties in the transmission of information and knowledge. Different groups develop their own frameworks and language for handling problems, and communication between these cultures can become difficult.

One way of reducing the effect of these difficulties, while reducing the time required to make decisions, is to use familiar solutions. Designers draw on experience (their own and that of others) to come up with a particular design solution for a specific site and a particular client. Relying on personal experience requires a considerable amount of knowledge of various solutions to problems, knowledge that is only acquired with experience. Young designers (with limited knowledge) and older designers (with greater knowledge) will, to greater or lesser extents, rely on solutions by others.

Solutions represented by office standard details and standard specifications represent the expert knowledge of the organisation that produced them. Another is the careful selection and use of pertinent and current information, i.e. simply limiting the amount of information available to the designer. This is done by sifting information as it arrives and rejecting a large proportion. The way in which this is done will be discussed in Chapter 8. Both strategies attempt to simplify the decision-making process.

Because designers will face design problems that are ill defined, poorly described or diffuse in nature, attempts must be made to define the problem clearly before it can be resolved. In some of the design literature, this process is described as 'questioning' (e.g. Potter, 1989). Definition of problems is made easier through the designer asking questions of himself or herself, and also of others. The aim of this questioning process is to be able to take full account of the information, explore possibilities and recognise the limitations, essentially a process of simplification.

Mackinder (1980) observed that the way in which architects make decisions is closely influenced by two factors; the amount of time available for the scheme to be fully produced (the most critical factor), and the overall importance of the material or product in relation to the total project. Also, where an architect's office was involved in projects where speed was of prime importance (industrial or commercial projects), there was a strong tendency to adhere to a well-established vocabulary of materials for structural forms, cladding materials and finishes. The more individual the project, the greater the amount of time was said to be required for research and development work. Mackinder concluded that projects involving any extensive use of new materials or systems (no previous knowledge or expertise) can only be entertained if time is available.

'Fitness for purpose'

Designers are perhaps better known for their creative endeavours than their practical abilities, yet when it comes to detailing and product selection, clients expect their professionals to act in a dependable and, above all, practical manner. The factors affecting choice of building product have always been its properties, or 'fitness for purpose', its cost and its availability. However, a number of other factors are now beginning to influence choice, some of which are dependent upon legislation. Such factors include the safety of the product (both during construction and in use), its estimated durability in use, its embodied energy and its environmental impact. In practice, some of the factors will conflict. Specifiers have to make a choice within the time and budget available and use their professional judgement. (Once chosen, the specifier can then confirm it in writing, see Chapter 5.)

Information provided by manufacturers, sometimes supplemented by verbal information from the trade representative, should enable specifiers to make an informed choice about which particular product to select. When asking designers what criteria they employ, they are likely to say that they are looking for products that are fit for purpose, i.e. that suit their particular requirements. Fitness for purpose can be broken down into a number of inter-related criteria for product selection and will include:

- aesthetic properties;
- maintenance characteristics;
- durability;
- technical characteristics (performance);
- environmental considerations;
- health and safety;
- security (vandalism);
- ease of handling on site (buildability).

All built structures will fail eventually, some sooner than others, although the end of a building's design life can be anticipated, and often extended, with sensitive and regular maintenance. It is the unexpected failures that cause the most concern for both designers (who may be liable) and owners (disruption and cost of putting it right). Failures of materials and/or execution can never be entirely eliminated, even with mass-production techniques and their associated quality-control procedures. Designers need to understand the limitations of the materials and manufactured products that they detail and specify, especially any changes in physical size due to moisture or temperature changes and the practicalities of working and/or fixing the materials on site if failure is to be minimised.

We cannot aspire to the theoretical perfection of Oliver Wendell Homes' wonderful one-hoss shay (The Deacon's Masterpiece). Tired of individual components wearing out at different times, the Deacon's Masterpiece was designed so that all the components lasted for the same length of time, until it all fell to pieces in a moment. However, the designer should be concerned that the life of individual components is commensurate with either the design life of the building or the life of its anticipated replacement. With increased emphasis on cost and value for money comes a concern about 'over-specification'. Given the cost associated with building components, product specification naturally forms a major part of any value-management exercise. Carried out before the drawings are sent out to tender, there may be pressure to substitute products with cheaper products. Under-specification will become evident when a component or system fails, but over-specification might only be evident when a building is finally demolished or substantially remodelled. Consideration of design life and service life at briefing and subsequent design and specification stages can help to eliminate over-specification, as can the early use of value management techniques.

Aesthetics, durability and performance

For designers, the aesthetic appeal of a product is often top of the selection criteria, especially where it is to be seen and experienced when the building is complete. Internal and external finishes are dependent upon the materials used and the manner in which they stand up to weathering and daily use. Aesthetics is very much a personal matter, with different specifiers having their favourite materials and manufacturers. Furthermore, many design offices develop a particular architectural style that may rely on the use of certain materials as an underlying approach to all their building projects. When dealing with an existing building, aesthetics takes on a different slant, especially when trying to match new materials to existing materials. Here, the specifier is constrained by the existing fabric, and so choice will be constrained in a different way.

The physical characteristics of components will determine the durability and performance of the building. As indicated above, design life and service life need to be considered at the specification stage. Specific information on durability is usually provided by the supplier of the product so that the specifier has to place a certain amount of trust in the information provided. However, durability may also be built into the common manufacturing standards and enshrined in national standards such as British Standards, DIN or ASTM standards. Although these may not actually define the life of the product, they may require properties that will ensure a satisfactory life, especially where this affects public safely. Consider, for example, flue liners that are attacked by flue gases and where failure allowing their escape would be hazardous. The durability of these may be indirectly controlled by specifying a minimum thickness.

With increased attention being given to both the design life and service life of buildings and their individual components, the durability and the anticipated performance of the materials in service need careful consideration. Frequency and ease of replacement are considerations for items that have a limited service life, as should be ease and cost of maintenance when selecting one product over another. It is becoming increasingly common for building-insurance organisations to lay down strict criteria that can influence choice of products, especially where they are concerned with asset and facilities management. Materials and components with short service lives or those that carry a greater degree of risk of failure may be effectively blacklisted.

Costs

It is not an easy task to establish the cost of some proprietary products, nor is it easy to separate out the cost of the product from the cost of the service provided by the manufacturer. This is an important point to make, simply because the service provided to the specifier will be different from that provided to the contractor. For example, the specifier will require technical details and possibly help with detailing and writing product-specific specification clauses, and the contractor will require information on delivery, costs and possibly assistance with issues relating to buildability on site.

Principal costs

There are three principal costs to be considered at the outset of a construction project, the initial building cost, the cost of the building in use, and the recovery cost.

Initial cost

Also referred to as the acquisition cost or the development cost, the initial cost covers the entire cost of creating, or remodelling, the building. For many clients, this is their primary, and often only, concern. Initial costs cover land/building acquisition costs, professional consultants' fees, the cost of the materials that comprise the completed building, and the cost of putting it all together. Cost reductions may be possible by selecting less expensive building materials and reducing the amount of time required to assemble them on site, but this assumes that these costs can be discovered.

Cost in use

Otherwise known as the running cost or operating cost, the cost in use is set by the decisions made at the briefing stage and the subsequent decisions made during the design and assembly phases: affected by the choice of materials and the soundness of the detailing. For many years, running costs were only given superficial attention at the design stage, although this has changed with the use of life-cycle costing techniques that help to highlight the link between design decisions and costs in use. Materials and components with long service lives do cost more than those not expected to last so long and designing to reduce both maintenance and running costs may result in an increase in the initial cost. However, over the longer term, say 15 years, it might cost the building owner less than the solution with lower initial cost. It is a question of balancing alternatives at the design stage and educating the client about building costs in use because many clients will need some encouragement to part with their money up front.

Recovery cost

There is a third cost that is rarely considered – the cost of demolition and materials recovery. This is partly because the client may well have sold the building (or died) long before the building is recycled and partly because such costs are traditionally associated with the initial cost of the future development. Again, this may be of little concern to the current client who is looking for short-term gain with minimal outlay. However, if we are to take environmental issues seriously, then the recycling potential and ease of demolition should be considered during the design phases and costed into the development budget.

Cost of individual products

The cost of proprietary products is not always known to the specifier at the time of selection. There are a number of reasons for this. First, manufacturers are often reluctant to give the cost of their products to the design team, for fear that the specifier will choose simply on price and not on value. Second, in the UK and commonwealth countries, where it is common to employ a quantity surveyor (QS), it is the QS who gets the cost information, not necessarily the specifier. The effect of this is that some designers show a lack of interest in the cost of building components. The actual cost of the product to the main contractor will be determined by the relationship that the contractor has with the supplier and/or builders' merchants and the level of discount provided by the merchant on certain materials.

There is of course, no such thing as a 'free lunch'. If drawings and specifications are provided free of charge by manufacturers (as discussed in Chapter 3), then the cost of this service is built into that of individual products. Companies offering this service provide added value for specifiers, although the client does not obviously benefit from this service. Guarantees and warranties offer a degree of comfort to the specifier, although such insurances are only valid whilst the company is still trading, and only valid if the product has been installed as stated. Furthermore, product-specific guarantees may be valid only if the product has been installed by an accredited installer. Some products should only be installed by a number of specially trained and certified installers, which should help to ensure quality workmanship. There may be an increase in cost for this, but experience shows that the finished work is of a higher standard, and less problems occur during the installation.

Availability

Availability should be checked directly with the manufacturer. This is especially important when materials, components or systems are being specified that take time to transport to the site and/or have to be manufactured to order. Availability can have major implications for the programming of the construction works, both for the designer and for the contractor. Checking availability at the specification stage can help to eliminate problems later in the contract. This is true of both prescriptive and performance specifications – there is little point in writing a specification for a particular element that takes 6 months to manufacture and deliver to site when dealing with a fast track project unless it can be pre-ordered and programmed for. Early involvement of project managers, specialist sub-contractors and suppliers will help to define clear and achievable programme deadlines. This can be of great assistance to the specification writer in helping to determine his or her choice of product(s). Availability, or rather the lack of it, is often used as an excuse by contractors for substituting products on site, as illustrated in the case studies.

Evidence in disputes

The specification of a building product that may fail in use is perceived as a major cause for concern by design practices, the majority of which go to great lengths to reduce their exposure by specifying products that are known to them. The selection of 'green products' represents a change in existing behaviour for the majority of designers and higher perceived risk. Mackinder (1980) noted that the partner of an architect's office influenced 'major' decisions regarding product selection, strongest at Stages C and D during which important decisions were taken that influenced the structure and the finished appearance of the building. As the design progressed to the detail design stages, where the number and variety of decisions increased, their importance in terms of both the cost of the job and the overall appearance of the building was seen to be reduced; Mackinder referred to these as 'minor' decisions. With the perceived reduction in the importance of the decision-making, it was common practice to involve less senior members of the office, often with varying degrees of supervision from their more experienced colleagues. There would appear to be a hierarchy within the office, with the recently qualified (younger) members of staff, with little experience, supervised by a more senior member of the office, who in turn was supervised where major product selections were concerned by the most senior member of the office, the partner. A combination of the specifier's experience and their position in the office hierarchy would appear to influence the relative importance, as viewed by the partner, of the products being selected.

In interviews with specifiers, their main concern with regard to risk was related to defects in the products they may select. However, in a comprehensive survey of construction problems involving specifications that led to claims, litigation or arbitration

(Nielson and Nielson, 1981) deficiency in product performance was some way down the list of problems. Top of the list were problems caused by the use of 'or equal' (25 per cent of cases). This was followed by ambiguous writing; differences between specifications and plans; problems regarding buildability; and inaccurate technical data (12 per cent each). Together, these five problem areas accounted for 73 per cent of the cases. Problems with product specifications accounted for only 8 per cent. Caution is required here because America has different specification habits to other countries, but the data are useful in helping to highlight some common problems faced by specifiers in other countries.

Specification writing

Specification writing is a skill, and the success of the specification will depend very much on the abilities of those involved in its production and the constraints under which they work. It may seem like a statement of the obvious to say that to be usable, written specifications must be easy to read, but they do not always have this virtue. Here, we look at the task of the specification writer and the constraints and controls under which he or she works. Guidance on writing style, the ability to accommodate changes, and the need to constantly check the written specification against the overall brief and design objectives is provided. The chapter concludes with some thoughts on the problem of staying up to date.

The specification writer

Once a decision has been made to use a particular product, the designer has to make reference to it in the contract documentation, i.e. it has to be 'specified'. This can be done by referring to the product on the drawings, by including it in schedules of work, and/or through the written specification. The written specification is an important document in ensuring a quality building for which adequate resources must be allocated by the office manager (see Chapter 6). The specification must be well written, be comprehensive and be free of errors, a task that requires considerable time. It is an essential part of the design process that requires particular skills in researching different product characteristics and being precise in communicating those requirements to a variety of organisations and individuals.

Specification writing is a skill. In the United States, there is a separate profession of specification writers, i.e. it is carried out by qualified specification writers who dedicate all of their time to the task. One can see from this that the task is seen to have greater importance, and the role carries greater prestige than it presently does in the UK. Some large offices in Britain employ people to write and check specifications, leaving the designers to design and the managers to manage. Such an arrangement calls for close co-ordination between designer and specification writer. However, the majority of British design practices are small, and so the specification is often written by the same person who carried out the design and detailing of the scheme, which means that they only spend a small amount of their time dealing with specification writing. In smaller offices, the designer has little option but to write his or her own specification. The point being made here is that the majority of specification writers are 'part-time', and therefore it is particularly difficult to be expert at this task. But expert they must be because there will be legal implications if the specification is wrong.

From the authors' experience of design management, it is clear that some people are well suited to specification writing, whereas others are not. Good specification writers tend to be individuals with a very good technical ability and considerable experience of both design and building operations. They should be precise and have an eye for detail. They also have to be exceptionally good at interpreting designers' drawings, i.e. they need to be aware of the project's design goals. They must also be able to communicate in writing. Because of their training and background, many designers prefer to use graphics to communicate in preference to written media and, as a result, may not be particularly adept at writing good specifications. Thus, those working in very small offices who have little option but to undertake this activity need to learn this skill. Ideally, specification writers need to have excellent knowledge of construction materials and products, construction methods and project management. They should also be aware that disputes arising from errors in the contract documentation will have serious legal consequences, so attention to detail is paramount. Once the requirements have been established, the next step is to express them clearly, concisely and in a logical manner in the written specification.

A job advertisement for a specification writer in *Building Design* (2000) helps to highlight some of the required characteristics. The advertisement made clear that for this 'key' position in their 'busy' office, they would require an individual with the following talents for this 'challenging opportunity':

1. 'Previous experience of writing specifications using NBS and office standard specifications in MS Word format in a PC environment' (implying that NBS is not used exclusively).
2. 'An excellent working knowledge of materials, construction methods and systems used in a range of buildings, both new build and refurbishment' (implying a certain level of experience and maturity).
3. 'Solid expertise in British, European and other international standards, codes of practice and legislation relating to the construction industry, along with the ability to interpret and implement regulatory and technical data'.

4 'The ability to work effectively as part of a team and communicate confidently at all levels' (we would have put this at the top of the list).

Writing the specification

The primary aim of the written specification is to convey information to the reader that cannot be easily indicated on the drawings and schedules. The contents of the document will be concerned with the quality of the materials and the quality of the workmanship, neither of which can be adequately shown on drawings. Although a large proportion of the document will be common to many different projects by many designers, specific details will be dependent upon the nature of the particular building being specified and the design philosophy of the designer or the design office. As noted above, writing the specification does involve considerable knowledge of construction, materials and working methods in addition to good writing skills if this practical document is to be of value to the reader. A guide to specification published in 1930 noted that a complete knowledge and understanding of the details of building construction, properties and cost of materials were paramount. Until this was mastered, it would not be possible to write a specification (Macey, 1930), a sentiment that holds true today.

Badly written specifications will result in claims for extras from the contractor. Willis and Willis (1991) claim that there are two essentials to specification writing: to know what is required and to be able to express such requirements clearly (echoing Macey's earlier advice). They note that many specifications fail because of shortcomings in the first stage. There may have been insufficient thought and/or insufficient knowledge of building construction.

1 Insufficient thought. To accuse professionals of insufficient thought may, on first sight, be a little unfair. As professionals, architects and other consultants have a duty of care to their clients and are expected to behave in a 'professional' manner. They should be capable of applying sufficient thought and sufficient knowledge to the problem in hand. However, insufficient thought may be given to a specification more because of a lack of adequate resources, especially time, rather than because of simple incompetence. It matters little how knowledgeable or how good an individual is if inadequate time is allowed to consider the options carefully, to make an informed decision and confirm this decision in writing within the specification.
2 Insufficient knowledge of building construction. At first sight, one might assume that insufficient knowledge is only a problem for those just starting work in the building industry. We noted earlier in the book just how little time is spent teaching architects the art of specification, and many observers of architectural education on

both sides of the Atlantic have become increasingly critical of the small amount of technical instruction within architectural programmes. Because of this, young practitioners have to learn through experience in the design office and thus require close supervision and support in their early years. However, problems of insufficient knowledge can also arise when an experienced designer is faced with a new type of building, a new problem or new products. Again, time is required for the individual to acquaint himself or herself with the new information before informed decisions can be made.

Naturally, to express requirements clearly in the specification, the writer first has to know what is required. Vague specifications are often an indication that the decision-making process had not been resolved at the time of writing. Bowyer (1985:11) sums it up rather neatly:

> Unless the designer knows what he wants he cannot expect either the specification writer to describe it, the estimator to price it or the builder to construct it.

Both of the essentials identified by Willis and Willis can be addressed through:

- adequate time to complete the task, i.e. good programming and management;
- easy access to current and relevant information;
- employment of experienced staff;
- close supervision of less experienced staff;
- continual education and training.

Adopting a systematic approach

The order in which the specification is written will be determined by the characteristics of the individual and any office procedures. Critical is the ability to approach the task in a logical and ordered manner, completing one section before moving onto the next. Preliminaries are often the last section completed because they are largely independent of the work sections. It is good practice to specify the work required before attempting to complete the materials and workmanship clauses, i.e. to identify what needs to be done and then what materials and methods should be used to achieve the task. Consistency is the key, and this is helped by using industry-standard formats. In situations where a non-standard format is used, it is essential that the document is set out in a logical manner, complete with an index of main headings to guide the reader.

An ability to accommodate design changes and information from others who are party to the design process is also required, in short, the ability to organise one's own work within the programmed time-scale. One way that specification writers can maximise their time and co-ordinate the input of others in the design process is through the use of master specifications. These reduce the costs associated with producing the written document.

Standard formats

Well-managed design offices recognise the importance of consistency, and the majority have developed standard formats for a variety of tasks, including the written specification. Standard formats can save an enormous amount of time, but care should be taken to ensure that errors are not being transferred from one project to the next through lazy copying without due care (see Chapter 6). Over time, different countries have developed their own 'standard' specifications as a means of guiding specifiers and also in an attempt to bring some form of consistency and common documentation to the building process, examples being America's CSI SPECTEXT, Australia's NATSPEC, the Netherland's STABU, and the UK's NBS. Although these are widely promoted and used, other specification packages are available from commercial suppliers, in addition to which large design organisations tend to use their own particular system. As a result, these 'national' systems have never become universally used.

There is a powerful argument for everyone in construction to use the same system, although this ideal does not seem attainable in the foreseeable future. National systems are a useful tool in ensuring clear, concise and (more importantly) familiar documentation for individual projects. In addition to their economic advantages to both designer and contractor (and hence client), there should be less likelihood of ambiguity and confusion caused by unfamiliar looking documentation. Assuming, for one moment, that everyone in construction could agree to adhere to a national format for specifications, this would form part of a quality-assurance package for the industry. One might be tempted to suggest that some of construction's persistent ills could be eased by such an approach. Perhaps, this is wishful thinking, and there are also arguments against this approach. National standards do tend to be (too?) complex, and the format and defaults may not be to everyone's taste. Time and skill are required to understand them and apply them correctly, so that they are just as open to misuse and abuse as other systems.

National Building Specification (NBS)

In the UK, the National Building Specification (NBS) is widely used. Available as computer software, it helps to make the writing of specifications relatively straightforward because prompts are given to assist the writer's memory. Despite the name, the NBS is not a national specification in the sense that it must be used: many design offices use their own particular hybrid specifications that suit them and their type of work.

NBS is only available via subscription to the service provider, which ensures that the user has a document that is kept up to date through regular revisions. At the time of writing, it is still available in paper format, although more recently, designers have realised the potential of completing the specification in electronic format, 'specification writer'. The NBS is an extensive document containing a library of clauses. These clauses are selected and/or deleted by the specifier and information added at the appropriate prompt to suit a particular project. The NBS ensures a consistent format, and the provision of prompts reduces the danger that some element is forgotten. However, as with all templates, the quality of the finished specification still depends upon the ability of the individual to fill in the gaps. The NBS is available in three different formats to suit the size of the particular project, ranging from Minor Works (small projects), to Intermediate, and Standard (large projects).

Specification language

Specifications do have their own language: a language that takes the inexperienced specifier and contractor some time to understand and become familiar with. However, there are guides, and standard formats can go a long way to achieving consistency. The CSI's *Manual of Practice* (CSI, 1992) provides comprehensive guidance on specification language. Consistency of word usage is key to a well-written document, and the writer must take care always to use the same word or phase in the same sense throughout the document if confusion of the reader is to be avoided. For example, one of the most common causes of confusion can arise when specifying sizes of components; all sizes should be either 'unfinished' or 'finished' sizes. Where it is necessary to use both conventions, it is important that the specifier makes it clear to the reader whether each size specified is 'finished' or otherwise.

Standards and Codes of Practice can, if used carefully, help to shorten the description of a particular material. The extent to which British (BS), European (EN) and International Standards (ISO) and Codes of Practice are included in a specification is often a point for debate. As discussed in Chapter 3, no one designer will have a working knowledge of all the relevant standards and codes; nor, for that matter, will the contractor. It is the specialist sub-contractors who will know and understand the standards and codes that apply to their particular area of expertise. Many of the standards have sub-divisions relevant to the same material, and so the specifier must make sure that the reference number quoted includes the correct sub-division. A good rule of thumb is to specify only standards that the specifier and/or the design office are familiar with. Unfortunately, it is common practice to quote general standards without taking sufficient care (because of a lack of time or simply through laziness) to ensure that it applies to the particular job, and the specification then becomes somewhat meaningless. This is a particular problem with rolling specifications from one job to the next and, to a lesser extent, with the office master specification, if not maintained on a regular basis (see Chapter 6).

Imperative or indicative mood

The manner in which a specification is written is important and can help in reducing repetitious and tedious sentence structures. Specifications can be written in the imperative mood or the indicative mood. The imperative is preferred for its clarity (CSI, 1992; Cox, 1994).

- Imperative mood. The imperative sentence is concise and easily understood because the verb that defines the action forms the first word in the sentence. The reader is directed by verbs such as Apply, Install, Remove, etc. The imperative mood is seen as the major factor in producing clear and concise text, makes it quicker to write (and read) and hence saves time and money. Example: 'Apply two coats of emulsion paint to ...'
- Indicative mood. The indicative mood is written in the passive voice. Sentences require the monotonous use of 'shall' and can be unnecessarily wordy. Example: 'Two coats of emulsion shall be applied to ...'

Style

Proper style will ensure clarity, brevity and accuracy. The style adopted should be consistent within the design office regardless of the degree of detail required for a particular project. Specification writers should use:

- short sentences;
- simple sentence structure;
- plain words and terms.

Certain phrases should not be used in specifications (or on drawings). A favourite for work to existing buildings is '... to match existing'. For the majority of existing buildings, directives such as 'brickwork to match existing' is impossible to comply with, simply because the exact brick is no longer produced, and so a brick that is a very close match is used. Moreover, such descriptions invariably forget to mention items that may not be apparent to the contractor. The specifier may be aware that lime mortar is used in the original work, but this need not be apparent to the contractor, who may assume the use of a cement mortar. This phrase is a quick and convenient way of not specifying.

Clarity, brevity and accuracy

The specification must be written in such a manner that it conveys the intentions of the designer to the contractor. This may appear to be an obvious statement, but specification writers must constantly bear in mind the fact that readers of the specification will not have been party to the decision-making process that led to the contract documentation. The readers can only read the documentation to see what is required of them. We mentioned earlier about the need to be able to express one's intentions clearly, for which sufficient thought and sufficient knowledge of construction are prerequisites. In many respects, the use of 'standard' specifications has helped individuals to write specifications because the format is already supplied. The writer then has the relatively simple task of deleting the clauses that do not apply and adding information as appropriate.

Designers have their own way of working, and many have 'golden' rules that they apply when designing, detailing and writing specifications. In offices where managerial control is not particularly good, this can and does lead to information taking a variety of slightly different forms, reflecting the idiosyncrasies of their authors. The end result can look unprofessional, can lead to confusion and, in the worst case, can result in errors on site. Professionally managed design offices take a much more considered and controlled approach. Designers work to office standards of graphic representation and to a standard approach to detailing, product selection and specification writing. Guidance for members of the design organisation is provided in the office quality manual. We suggest six 'golden rules':

1 Clarity and brevity. The most effective information has clarity and is concise. This is far easier to state than to achieve because it is impossible to represent everything in an individual's mind on a drawing or in text. The skill is to convey only that which has relevance and hence value to the intended receiver. This can be a matter of knowing when to stop writing. This will help the receiver to avoid information overload and enable him to concentrate on the relevant information without unnecessary distraction. Describe items once and in the correct place. Repetition should be avoided.

2 Accuracy. It is important to be accurate in describing requirements because confusion will lead to delay and errors on site. Use correct words to convey exact instructions, use correct grammar, units and symbols, and avoid ambiguity. Words and symbols should be used for a precise meaning and be used consistently for that meaning throughout the document. Instructions should be given accurately and precisely. Use a limited vocabulary of words. The document should be complete: do not leave out important information or leave clauses partially completed.
3 Consistency. Whatever the approach adopted by the design office and the individuals within it, it is important to be consistent, both in the meaning of words and in the approach to specification decisions. For example, if specifiers do have an individual and different approach to detailing, those on the receiving end should be able to interpret instructions as long as the approach remains the same. Use of graphics, dimensions and annotation should be reassuringly consistent across the whole of the contract documentation. CAD packages and the use of the CI/SfB should both help to achieve this goal.
4 Avoiding repetition. Repetition of information in different documents is unnecessary, is wasteful of resources and, when repeated slightly differently (which it invariably is), can lead to confusion. Repetition, whether by error or through an intention to help the reader, must be avoided both within and between different media. Eliminate unnecessary words and sentences in the written specification and avoid notes on drawings wherever possible. Be concise.
5 Redundancy. There is always a danger that superfluous or redundant material will be included in a project specification. Text from the master specification may be redundant because it is not relevant to a particular project. Rolling specifications from one project to the next invariably result in redundant text. A favourite example of the authors comes from a large refurbishment project where the specification said 'Remove defective render ...' There was no render on the project. In this example, the specification had been rolled from another project that did have rendered walls. The document not only becomes larger than it should be but will lead to confusion and may well undermine the credibility of the written specification (and those who contributed to it). Careful editing should help to remove the majority of redundant material.
6 Checking. Check and double check for compliance with current codes and standards, manufacturers' recommendations, other consultants' details and compatibility with the overall design philosophy. Common problems encountered by site personnel can be reduced significantly through a thorough check before information is issued to the contractor, and not too long ago, it was common for drawing offices to employ someone to check all drawings and specifications before they were released from the office. Unfortunately, in the constant drive for efficiency

and ever-tighter deadlines for the production of information, such checks have been left to the individuals producing the information. Self-checking is suspect and subject to error simply because of the originator's over-familiarity with the material. Managerial control is essential in this regard and must be costed into fee agreements. Checking for omissions and errors, accommodating design changes and auditing the process through quality management systems can save time and confusion.

Typical specification formats

Building specifications, whether for new or alteration works, have two main types of clauses. First are those that describe the general conditions under which the work should be carried out and various obligations of employer, contractor and designer. Second are those that describe the materials and workmanship required in detail. The contents of a typical specification are grouped under common arrangement work headings from A to Z:

A Preliminaries/general conditions
B Complete buildings
C Demolition/alteration/renovation
D Groundwork
E In-situ concrete/large precast concrete
F Masonry
G Structural/carcassing metal/timber
H Cladding/covering
J Waterproofing
K Linings/sheathing/dry partitioning
L Windows/doors/stairs
M Surface finishes
N Furniture/equipment
P Building fabric sundries
Q Paving/planting/fencing/site furniture
R Disposal systems
S Piped supply systems
T Mechanical heating/cooling/refrigeration systems
U Ventilation/air conditioning systems
V Electrical supply/power/lighting systems
W Communications/security/control systems
X Transport systems

Y Services reference specification
Z Building fabric reference specification.

Preliminaries – the general conditions

At the front of a specification is the preliminaries section that deals with the general conditions applying to a particular project. These range from the project particulars (A10), schedule of drawings (A11), site description (A12), a brief description of the work (A13) and the contract (A20) through to the management of the work (A32), quality standards (A33), to safety (A34), specific limitations (A35), etc.

Materials and workmanship clauses

The materials and workmanship clauses are important because it is here that quality standards can be clearly stated. Care should be taken to ensure that individual clauses are both relevant and precisely worded. In addition to the quality of the materials to be used on site, the specifier will also be concerned with the quality of the workmanship. The need for good workmanship and safe working practices is self-evident and must be specified correctly. The specification writing packages provide a very useful list of all clauses, and the challenge for the specifier is to decide which to delete and which to keep and complete in full to match the requirements for the project.

Specifications for new works

This section forms the main body of the specification and should be set out as illustrated in the typical list of contents, described above.

Specifications for alteration work

When dealing with alterations to existing buildings, it is sometimes easier to subdivide the specification according to certain areas of the building and/or according to particular rooms. By composing the specification in this manner, everything connected with a particular area is described together and is generally preferred by the workmen on site. On larger alteration projects, it may be easier and more efficient to use a mixture of both methods, with new work classified by work section and alteration work by room. The layout of the specification for alteration work will depend upon the size and complexity of the job and the format adopted by individual design offices to best suit individual projects.

Specifications for conservation work

The problem of clauses like 'to match the original' has already been discussed briefly. The specific problem facing the specifier of conservation work is that both the skills and the materials used in the past might no longer be commonly available. In such cases, the specifier may need to include descriptions of operations that would otherwise be unnecessary. For example, when brickwork is laid with cement mortar, the joints are struck as the bricks are laid. If lime mortar is used, this must be carried out as a separate operation. Rather than referring to modern standards and specifications, the specifier may wish to refer to those in earlier trade manuals. A useful source here is the annual *Specification*, which provides specification clauses used in work from the end of the 19th century.

Prime costs and provisional sums

In situations where it has not been possible to define everything to be specified, the designer can include prime costs and provisional sums. Prime cost (PC) sums will be included in the tender for goods to be obtained from a nominated supplier. This sum will be adjusted against the actual cost of the products selected. For example, bathroom suites or similar items are often included in the specification as a prime cost simply because the client has not yet decided what style and colour are required. Provisional sums are used to cover work and/or items for which insufficient information is available at the tender stage and which cannot be measured or priced accurately, as such provisional sums are particularly useful for alteration works. Another use for provisional sums is to cover the work to be carried out by statutory authorities and utilities companies, for example the connection of mains drainage, water, gas and electricity.

Contingency sum

Building projects will, to lesser or greater extents, have a number of unknown elements that only become known as the work on site progresses. Typical examples are unexpected difficulties with ground conditions and unexpected problems when existing buildings are opened up. The contingency sum is essentially an undefined provisional sum of money to be used if required.

Staying up to date – a constant challenge

Professionals have a duty to stay up to date with current regulations and codes, current building practices, changes to forms of contract, and developments in materials and products, both new products and those rendered obsolete. Although this may sound a relatively easy thing to do, in practice, it presents a series of challenges. One of the problems facing many designers is that they do not need to access the information sources all the time. Indeed, many designers work on many different stages of jobs, and the physical act of specification writing does not take up a great deal of their time. This is particularly so of designers who run a project from inception to completion, mainly those self-employed and working in small offices and/or on small projects. For these individuals, access to information to assist them with the specification writing may only be required every 12 months or so, and then only for a few weeks until the task is complete: they therefore need a reliable and current source of information that can be accessed quickly.

Organisations that subscribe to one of the information providers, such as the Barbour Index or NBS, are kept up to date with the majority of new developments that affect specification writing. If not, they will have to rely on reading about changes in the professional journals and trade information. In practice, the majority of practitioners try to stay up to date through information from a variety of sources. Knowing where to find a particular piece of information in a crisis is one of the pre-requisites of staying in a job. More specifically, the specifier needs to keep up to date with:

- Building regulations and codes. Subscription to one of the on-line information providers will ensure that the regulations and codes accessed during the design and specification process are current.
- Building practices. Staying up to date with current building practices is a little more problematic. In part, this is because it is rare for two designers to agree on the best way of detailing and specifying a building.
- New materials and products. Manufacturers are constantly seeking to improve their products and expand their market share; thus, products may be 'improved' or replaced as part of their strategy. Specifiers have to keep up to date with these developments in order to specify effectively. Another problem is keeping up to date with materials and products that may no longer be available, perhaps because of safety concerns or simply because the product was not commercially viable.

Existing buildings

For designers dealing with existing buildings, the challenge is reversed. Their problem is to find the relevant codes and building practices that existed at the time the building was constructed (and or re-modelled). In order to ensure satisfactory performance of the structure and building envelope, it may also be necessary to match the properties of the new work with that of the existing. Yeomans (1997) has clearly demonstrated the difficulties for the practitioner in finding information on early building products. It may be necessary to either carry out a search of early literature or undertake tests to determine the properties of the existing materials before the specification for new work can be written.

Managing the specification process

However small or large the project, there is a need to manage the specification process. Office policy and procedure will influence the way in which design and production information is produced. A consistent approach is required, and quality-management systems may be seen as an essential tool in ensuring consistency. Specifications take time to produce, and therefore, resource allocation and the effective use of computers need careful consideration. The management of other consultants' contributions, as well as the need to manage design changes and product substitution, is addressed. The chapter concludes by looking at the issue of 'new' products from the viewpoint of the design office.

The specifier's milieu

The environment in which the specifier works will, to a greater or lesser extent, influence the decision-making process that is central to the specification process. Organisational culture is an integral part of the specifier's daily environment and will influence how individuals behave in the design office (Emmitt, 1999). This culture comprises the collective experience of the office, its possible use of standard details and the sources of information available in the office. All this, as well as the previous experience of the specifier, will influence the specification process. So, too, will size, market orientation, service provision and the way in which the office is managed (Figure 6.1).

The specifier's office

Design organisations vary in size, have differing workloads, different staffing levels and hierarchies and often differ in the type of services that they offer. Each organisation will

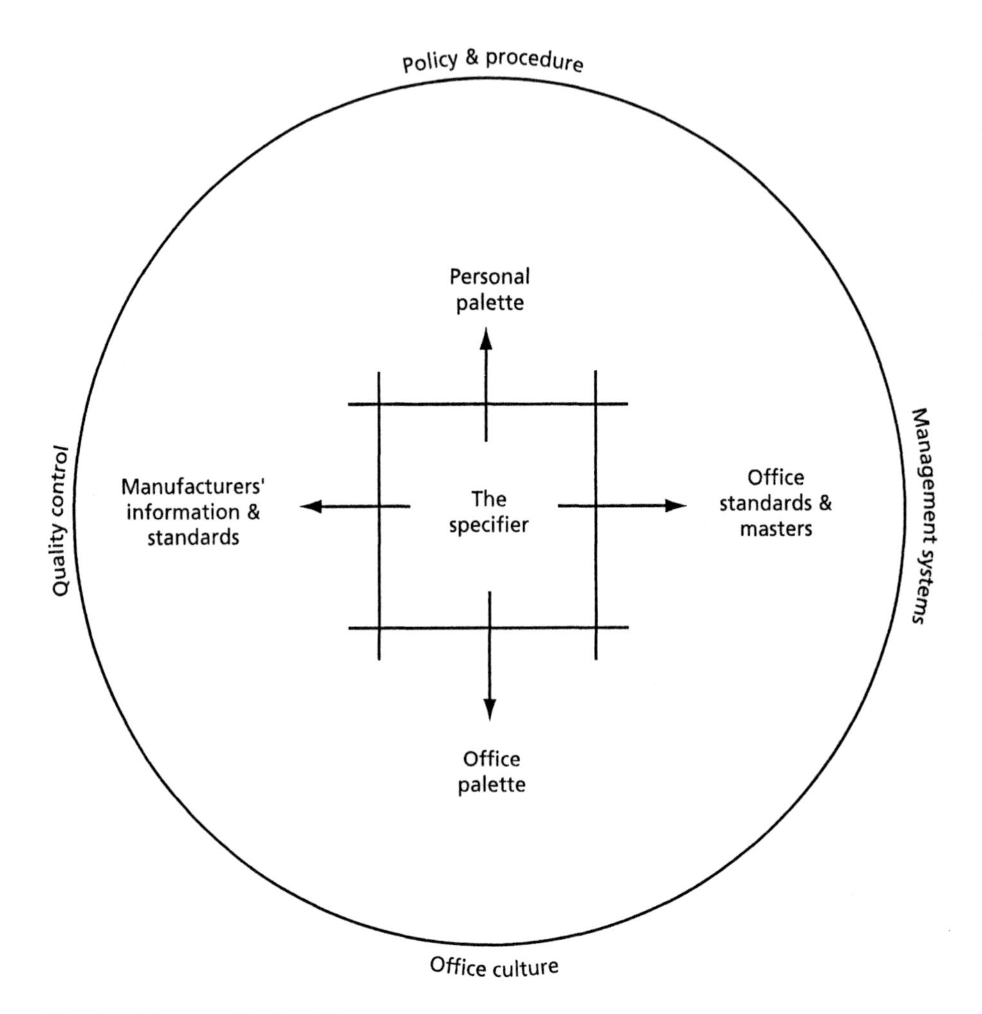

Fig. 6.1 The specifier's milieu

have its own identity, either by design or by accident. Office size is also a determining factor, but it is difficult to make international comparisons here. Although similar terms are used to distinguish different sizes of office, the divisions between large and small are made at different points. In the US, the division between small and large architectural offices has been made at the 20-person level, with Gutman (1988) noting that 90 per cent of offices are small. By comparison, this would also include offices that are defined as medium-sized practices in the UK, a category that comprises 85 per cent of offices. The implication here is that many specifiers working in this large proportion of offices are also engaged on other activities during their working day.

Design organisations also differ in the type of service they offer clients and the type of work in which they are engaged, which is more difficult to quantify. Allinson (1993) and Gutman (1988) have categorised architectural practices by the type of service they provide, dividing them into 'strong idea firms', 'strong service firms' and the 'strong delivery firms'. From this, it is reasonable to assume that the type of architectural practice may influence their attitude to the selection of building products. Symes et al. (1995) also reported that 80 per cent of the firms surveyed specialised in at least one building type such as housing or commercial and industrial buildings. Clearly, such specialisation will influence the range of building products that are selected and presumably the skill with which they are able to specify them. Offices will have more experience of products aimed at building types in which they specialise.

Control

Regardless of size or orientation, every design office needs to have someone in control. This person is usually the senior partner or managing director of the office. In large organisations, the manager may be the design manager, someone who takes responsibility for design decisions made by his or her design team. These managers will influence the behaviour of individual specifiers through general policy decisions, individual project management and the day-to-day design office management. Managers will also influence the process through their managerial style, be it autocratic or democratic. Control can be divided into three levels: policy decisions, individual project control, and day-to-day managerial control.

1 Policy decisions. Policy determines how the office is managed and how the specification process is controlled. The use of quality-management systems, master specifications, preference for performance or prescriptive specifications, etc. will colour the behaviour of specifiers in the office.
2 Individual project control. This will be tailored to suit individual clients and the characteristics of the design task. Quality parameters will be set out in the project quality plan.
3 Day-to-day managerial control. Management of individuals within a design office varies widely, from leaving specifiers to make their own decisions with minimal input from their manager, to very tight control where decisions are closely monitored, and approval is required from the design (or technical) manager for the slightest variation in office procedure.

Don't blame the specifier

The task of specifying is difficult to define since it continues, with varying degrees of frequency, throughout the design and construction phases. Many of the actions that the specifier goes through are, in the main, subtle and difficult to observe. As a result, the process may be difficult to manage unless it is fully understood and the implications of decisions taken recognised. Getting it wrong is costly, and therefore, adequate systems need to be in place to prevent mistakes extending beyond the office boundary.

When things do go wrong, the tendency of senior managers is to blame the specifier. This is a little unfair for a number of reasons, not least because of the complexity of the specifiers task. First, the organisation should have employed staff who are competent and then allocated them to tasks appropriate to their levels of experience and knowledge. If inexperienced, then they should be adequately supervised. Second, the managerial systems should be designed and put in place to prevent errors. Quality management systems and office quality-control checking procedures are essential tools, as are regularly maintained and updated office standards. When mistakes are found, it is essential that the office procedures are checked and then specifiers' actions checked for conformity with these proceedures. Once analysis of the process is complete, the system(s) can be revised to prevent similar mistakes happening again.

Office standards

Standard formats can be an effective tool in the quest for consistency of service provision. Standards represent an excellent knowledge base from which to detail familiar buildings, and many organisations try to prevent employees who leave from taking their 'knowledge' with them to a competitor. Effective use of such standards offers a number of benefits, but there are also a number of pitfalls to be avoided.

1 Advantages

- Quality control. Because standard details and specifications have been tried and tested by the design office over a number of years, they should be relatively error-free. They will have evolved to suit changes in regulations and to accommodate feedback from site. They provide consistency where there is some turnover of staff and a pool of experience to guide younger, inexperienced staff. Because they are familiar, tried and tested, standard details represent an effective means of quality control when applied correctly. Checked and updated at regular intervals, standards may contribute to the quality-management system of the office, reassuring clients and practice principals alike.
- Managerial control. The use of standard details and specifications can save the design office time and money because common details and clauses do not need to be reworked, merely selected from the design organisation's knowledge base. With increased downward pressure on professional fees, the use of office standards can help to ensure that the commission is profitable. Indeed, there may be

little time available to investigate alternatives.

- Risk management. The use of tried and tested specification clauses helps to limit the organisation's exposure to risk. Essentially, it is a conservative, or 'safe', approach to design.
- Benchmarking. When faced with an unusual detailing problem, the standards form a convenient benchmark from which to develop the detail and help to evaluate its anticipated performance in the completed building.

2 Disadvantages

- Perpetuated errors. Where errors exist in standards details and specifications (and they do), the errors are perpetuated through re-use on many projects until such time that the error manifests itself, sometimes after a long period of time. Unless careful checking and updating are undertaken, the use of standards can prove a dangerous habit.
- Incorrect application. Inexperienced members of the design office are often left to apply standards with little or no supervision. There is then a real risk that they may apply details incorrectly, and managerial control is essential if costly errors are to be avoided. Auditing the specification process is important in tracing specification decisions and identifying areas for improvement.

Feedback

One of the biggest complaints levelled at designers, especially architects, is their inability to return to projects in order to gather feedback. An associated problem is their reluctance to analyse projects to see what went right and what could have been better, with resultant information helping to inform future projects. Where the office uses a master specification, incorporation of feedback is essential for ensuring that the documentation retains its currency. We noted above the importance of taking feedback at the end of the project, analysing it, and incorporating relevant changes into the master specification as soon as possible. This needs to be done systematically and will be covered in any quality assurance systems implemented by the design organisation. There is another, more holistic, approach to feedback, and that is the ability to reflect during the process and learn from it (see Chapter 11).

Quality matters

In well-managed organisations, the writing of the specification and any alterations to it will be covered by quality assurance and quality control procedures that should be designed to eliminate errors and omissions. However, the management of many design organisations fails to live up to this ideal, with quality standards determined by the whim of the senior partner rather than by any written documentation. What is too frequently not realised is that this is an important factor in attaining and maintaining a competitive advantage. Information for building design is produced and used by organisations who are in business to make a profit. Organisations, regardless of size or market orientation, must give their clients, i.e their customers, confidence in the service that they provide. Those involved in the design and construction of buildings must also satisfy their clients with the quality of the finished building. There are two related issues, that of quality control and that of quality assurance.

Quality control

Quality control (QC) is a managerial tool that ensures that work conforms to predetermined performance specifications. Although quality control has been developed by, and is still very much associated with, manufacturing, it has been adopted more recently by the service industries as part of a quality management system. For manufacturers, the achievement of quality production, with long runs of products in a controlled and stable environment, is achievable and, assuming that the technology and personnel are correctly deployed, easy to maintain at a constant standard. When it comes to achieving quality control on a building site, the parameters are different. First, much of the work is carried out without the benefit of shelter from the weather, and emphasis on programming work to achieve a weather-proof envelope as early as possible in the assembly process is a prime concern. Second, the number of different operatives present on site at any one time (sub-contractors and sub-sub-contractors) makes the monitoring of quality particularly time-consuming for both the construction manager and the clerk of works. Work can be completed and covered up without anyone other than those responsible for its building knowing whether or not it complies with the standards set out in the specification.

To combat this, there have been various initiatives to move as much production as possible to the factory by way of prefabrication, leaving only the assembly to be done on site, that have been tried over the years with varying degrees of success. The effect has not always been an improvement in the quality of construction (Yeomans, 1988). For professional service firms, such as architects and engineers, quality control is more concerned with checking documentation against predetermined standards. Checking drawings, specifications and associated documentation before issue and the checking of other consultants' documentation for consistency with the overall design concept will help to ensure the quality of the information provided to the builder. Quality control is also achieved by adherence to current codes, standards and regulations.

Quality assurance

Quality assurance (QA) is a formally implemented management system that is certified, and constantly monitored, by an independent body, such as the BSI, to ensure compliance with the ISO 9000 series. QA is a managerial system that states what an organisation will do (in documentation), doing so by defining set procedures, and proving that they have been carried out. The process is designed to give the customer a degree of confidence that the promised standard of service will be delivered. Quality management evolved from early work on quality control in the American manufacturing industry, but it was the Japanese who took quality management to new heights. From the 1950s, they contributed to the Japanese revolution in continuous quality improvement, a revolution that has spread world-wide. Widely adopted in manufacturing, quality management systems have taken longer to gain widespread acceptance in the building industry, although many contractors and professionals now have certified quality management systems in place or claim to be working towards QA. In attempting to constantly please the client through a total quality management (TQM) philosophy, a step-by-step approach to continuous improvement, known as *Kaizen* in Japan, TQM has gained widespread acceptance. It is a people-focused management concept – a soft management tool – engendering pride in one's work and the desire to improve upon past success.

Quality of the finished artefact

Quality controls, the use of quality management, and the adoption of a total approach to quality from everyone involved in the construction process will be instrumental in determining the finished quality of the building. Quality for each individual building project will be determined by the following constant variables:

- assembly of the design and construction teams (procurement route);
- effectiveness of the briefing process;
- effectiveness of the design decision-making process;
- effectiveness of communication of information from designer to assembler;
- effectiveness of the assembly process;
- time constraints;
- financial constraints.

Because so much effort is often expended on the design and construction process, it is crucial to understand quality from the perspective of the building users. While space and facilities are critical to ensuring user satisfaction, so to are the materials that form the finishes. Quality materials and craftsmanship carry a higher initial cost than cheaper and (possibly) less durable options, yet the overall feel of the building and its long-term durability may be considerably improved.

The master specification

Specifying is essentially a task undertaken by individuals within an organisational framework. For the individual, the specification process is a research function from which decisions are made and communicated to others. To operate efficiently and effectively, the specifier needs access to relevant literature and to tools, such as the master specification, to help him or her complete the task with the minimum of effort in the time available.

The master specification is essentially a library of specification clauses used by the design office on previous occasions that have been assessed for technical suitability, filtered, co-ordinated and updated on a regular basis. It is not to be confused with rolling specifications from job to job. It is a vital part of the design organisation's expert knowledge system. Maintained and updated on a regular basis, the master specification can save individual specifiers considerable time and effort by reducing repetitive tasks. Correctly managed, over time, the master specification will help to ensure consistency because all project specifications are drawn from it. It will maintain and improve quality through feedback of good and bad experiences, help to keep the cost of production down, and aid the co-ordination of information. Thus, the master specification is a crucial resource for helping to ensure quality control and also providing a quality assured service to clients. The more effective and easier to use the master specification is, the greater the potential efficiencies and hence profit for the design organisation. This source of information is a valuable resource and must be managed accordingly if it is to remain of use to specifiers in the office (Figure 6.2).

Controlling the master specification

Someone within the design office, the chief specification writer, must be made responsible for the master specification. This person's task is to keep the document up to date and record all changes made to it in accordance with the organisation's quality management system. In small offices, this will be undertaken by a designer, or technologist, in addition to their other duties, and time must be properly allocated to the task. In larger offices, the chief specification writers will spend all of their time on specification matters and may well have assistance from other technical staff within the office in order to collect and analyse technical and product information.

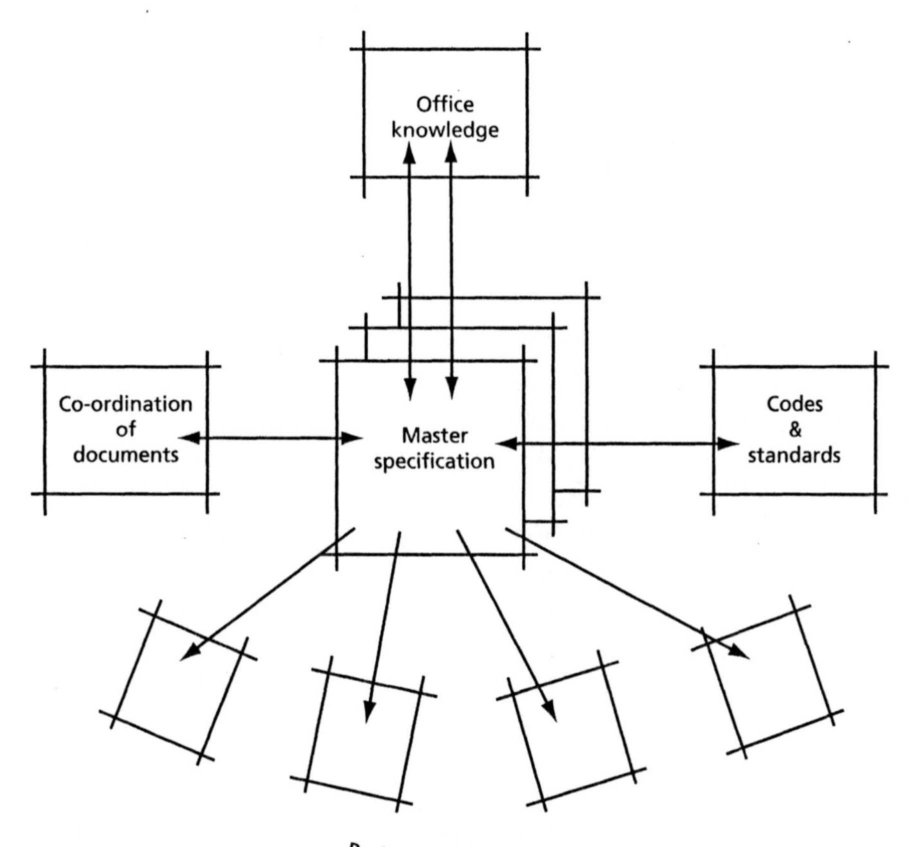

Fig. 6.2 The master specification

Because co-ordination is essential, it is common for design offices to set up the master specification based on a standard model (e.g. the NBS) and stick to that. Thus, initial choice of commercial master is an important managerial decision.

A key to efficacy and quality is the ability of the individual in charge of the master specification to maintain it. Done well, all specifiers and their project specifications will benefit. Done badly, all specifiers and their project specifications will suffer. This means that an essential characteristic of the chief specification writer is the ability to keep up to date with new products and with new codes and standards. A greater challenge is to keep up to date with technical changes to existing products, with changes in methods of construction, and with revisions to standards and codes when they are issued. This person also has to have an awareness of contractual issues and legal liability. Clearly, such changes may necessitate revisions to the master specification, which must be the responsibility of the chief specification writer. All changes to the master specification should be recorded and their potential implications noted and communicated to the specifiers in the design office to avoid abortive work. Only the chief specification writer should be allowed to make changes to the master. Given the importance of the document, some design offices operate a double-checking system where the document is checked by someone other than the chief specification writer. Although time-consuming, it is good office practice.

'Rolling' specifications

Care is required to distinguish the carefully and regularly updated 'master specification' from the 'rolling' specification. Rolling specifications are documents that have been used for a previous project and are simply rolled forward and adjusted to suit the next one. Their use is widespread but should be avoided because there is serious danger of including text that is inappropriate, and excluding that which should be included. Over time, other dangers such as references to superseded standards and discontinued products is a real possibility. This invariably leads to queries on site, over and/or under-ordering of materials, additional costs and claims. It is a lazy and potentially hazardous approach to writing project specifications.

By its very nature, rolling a specification from one project to the next is 'convenient' when time and other resources are at a premium, i.e. the specifier has not been allocated enough time to complete the task. Because the work is being rushed, problems occur largely because of inadequate checking of the text and inadequate checking against other project information such as detailed drawings and schedules. In the worst examples, site personnel become so frustrated with inconsistencies between the specification and other project information that they simply stop reading the specification. (One is tempted to ask: why bother writing one in the first place?) This leads to the dangers of changes on site and inappropriate levels of quality of materials, implications on cost and programme, and the enhanced risk of claims being made against the design office. Because of these problems, their use is not recommended, and well-managed offices have managerial systems that prevent their use. Individual projects require individual project specifications, based on the collective knowledge of the design office, not on the idiosyncrasies of the specifier.

Project specifications

At this point, it is appropriate to summarise some of the points made above.

1 Individual projects will require their own bespoke specifications. Just as every site is different from the next, so too are buildings and the specification.

2 The quality of the written project specification will be determined by office policy and the abilities of the individual doing the writing. In offices where different designers write their own specifications, there may be a wide variety in quality.
3 The use of quality management systems and reliance on the office master (and or a national standard format) can help to make the specifications more consistent within the office.
4 Rolling specifications should be avoided because of the dangers of erroneous or irrelevant clauses.
5 Specifiers should be able to use a master specification as their starting point, in the confidence that it is up to date and free of errors.
6 The establishment and, as importantly, maintenance of a master specification require investment by the office.

The degree to which individual projects draw on the office master specification will be determined by a project's characteristics and its compatibility with the master. Problems may be experienced in situations where a design office concentrates on, for example, residential work, and then acquires a commission for an industrial unit. In such situations, the master specification will be of less use than with the normal housing projects, and greater care is needed in its application to this unfamiliar situation. Decisions about product selection may be referred back to the office manager for approval. As noted in later chapters, great care (and adequate time) is required when dealing with products and information new to the specifier.

Specifiers have different ways of working, but it is regarded as good practice to develop and build up the specification as the detail design proceeds. Once the first draft has been completed, it is then a case of editing the document to suit final design decisions. As with the master specification, the project specification should be checked by someone other than the specification writer. Sometimes, this job is done by the design manager, sometimes by the chief specification writer. It is poor managerial control to issue specifications and drawings without a comprehensive check for errors and co-ordination. If the master specification is kept up to date as changes to materials and codes occur, then there should be no need for feedback from individual specifications. However, good and bad experience of materials, products and working practices gained from individual jobs should be considered and the master document revised to accommodate new knowledge.

The writing process

A systematic approach to project specification writing is recommended. Figure 6.3 illustrates the three main phases and the steps to be followed by the specifier. This

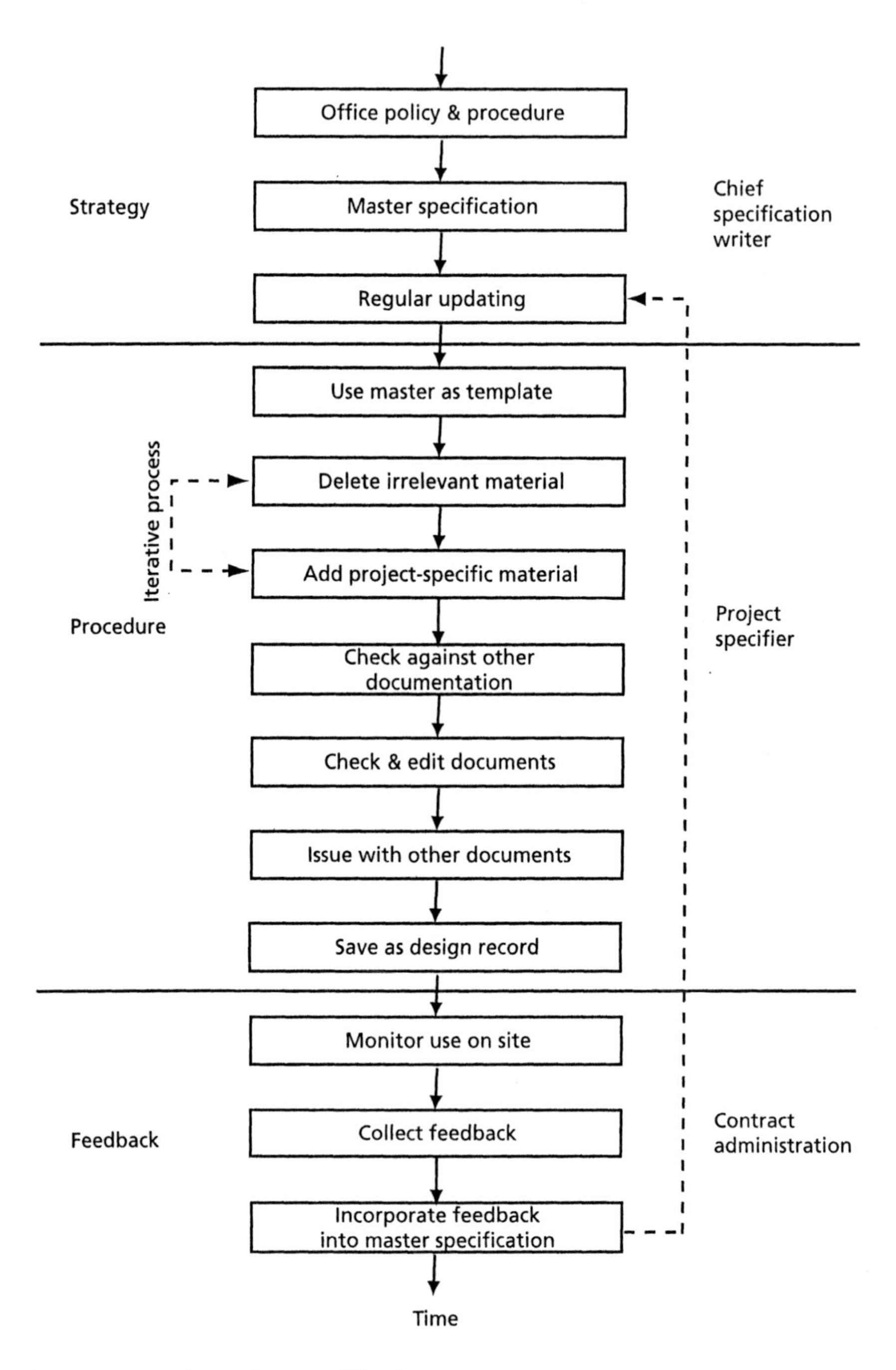

Fig. 6.3 Stages in project specifications

starts with the overall strategy, followed by the writing procedure and concludes with feedback. It is acknowledged that these steps may not necessarily be followed in a strict order, but the diagram should provide useful guidance.

Resource allocations

Given the importance of specification writing, it is surprising how little time is allocated to it in many design offices. In the less well-managed offices, earlier stages might have exceeded their time allocation so that later stages, perhaps seen as less creative, have to be squeezed if the office is to deliver the product information on time. The production drawings, and especially the specification writing, being at the end of the process are the two stages that frequently lose allocated time, resulting in rushed work that is inadequately thought through or checked. The result can thus be a document with too many omissions and errors that inevitably provide the contractor with opportunities for claims and/or inadequate work. Time and cost are closely related, and the manner in which these two valuable resources are managed will affect the quality of the service provision and that of the finished building. Clients want a quality building for as little financial outlay as possible, and (of course) they want it delivered in a short time period. From the designers' perspective, the budget is never quite generous enough to allow good-quality materials, and the time frame to achieve a good design is always too tight. Builders are then on the receiving end of cost-cutting exercises and tight programmes.

With careful planning and good managerial control, the majority of projects are delivered on time and within budget, but when things do go wrong, this invariably leads to the need for additional time and/or additional expense. It may be true that 1 or 2 weeks in the design and construction stages is negligible in the overall life of a building, which may be 100 years, but this argument should be made at the briefing stage, not when the project is starting to run behind schedule. The point is that good management begins with the correct estimation of the resources and time required at the beginning of the work and the discussion of this with the client at the briefing stage.

We have seen from the above that when allocating resources for a project, the design manager must allow adequate time for writing and checking the specification prior to issue. The task of specification writing should be clearly separated from the task of producing the production drawings. In practice, the tasks of detailing and specifying are often difficult to separate, but they are quite different tasks and must be costed accordingly. To produce a set of comprehensive, error-free drawings takes time, so does the writing of a comprehensive, error-free specification. They are inter-related, but separate, tasks and must be resourced accordingly even when draft specification clauses are written as the design proceeds.

Resource allocation takes on an even greater importance when dealing with existing buildings. Even where extensive investigations have been carried out, it is unlikely that precise requirements can be established until the building is opened up, i.e. as the work proceeds. Allowance will have to be made for changes to the specification (covered by the contingency sum), and the design manager must allocate sufficient time for the specifier to deal with such changes (covered by some contingency in the programme). This is important, because the time pressures placed on the specifier to make a decision may be more critical than on new build projects. Our experience of repair and rehabilitation projects is that clients are often reluctant to allow sufficient opening up of the building until the last moment, primarily because they wish to keep the building in use (health and safety concerns permitting) while the work proceeds. Again, this may exert additional time pressures.

Staffing requirements

In an attempt to be cost-effective, and hence competitive in the market for professional services, there is a tendency to use the cheapest available person for specifying. However, the cheapest person available does not necessarily mean the cheapest final project because staff with lower hourly rates tend to be those who are less experienced and, we would argue, not suited to writing specifications. From a manager's perspective, it may be useful to consider staff in terms of their experience, rather than their cost per hour (which may not necessarily be comparable).

1 Inexperienced staff. These are usually students or the recently qualified who are the cheapest resource in staffing terms. However, the need for constant nurturing and supervision makes the true cost of this resource considerably higher than it may appear from a balance sheet. A considered mix of advice from experienced colleagues combined with an ability to question conventional wisdom is desirable.
2 Experienced staff. Experienced staff are a design organisation's greatest asset. Capable of working with minimal supervision within the office managerial system, they can often produce accurate work fairly rapidly.
3 Over-experienced staff. Care should be taken to ensure that experienced staff stay up to date with current developments and do not rely entirely on over-familiar (and rarely challenged) solutions. Re-allocation of duties usually dispels any complacency very quickly.

Of course what usually happens in a design office is that the design manager has to use the staff available at the time (those who are least busy) rather than those best suited to the job. This can be avoided if the office is managed using the sequential system

(Sharp, 1991; Emmitt, 1999) where the job is passed along the supply chain, a systematic approach that can be very cost-effective if well managed.

Time and professional fees

Time is the most precious resource and the one that no one ever appears to have enough of. No matter what the task, we would all like longer to complete it (or do it better): there is nothing unusual in this. Time has an economic value, and for commercial concerns, the sooner clients receive their building the greater the financial return. Similarly, building designers and builders able to minimise the amount of time required to assemble a building, from inception to occupancy by the client, have a competitive advantage over those who cannot: a service many clients are willing to pay a premium for. To do this requires extensive knowledge of design, manufacturing, assembly and managerial skills.

Amongst other factors, the effective management of the specification process is key to ensuring a quality service and a quality product in the agreed timescale and for the agreed fees. This is an area that authors of books on office management and specification writing have avoided, but it is covered particularly well in Chapter 3 of the NATSPEC guide. Here, Gelder (1995) suggests that the time required to write specifications varies from 15 per cent for smaller projects down to around 2.5 per cent for the largest. No evidence is provided to back up these figures, and the author acknowledges that the figures are a 'crude' guide. However, if correct, the implications are that the smaller offices that are handling such small projects have the most to gain by a careful consideration of how they might improve the efficiency of this process. The actual time taken depends on a number of factors including access to available sources, expertise of the writer, thoroughness, timeliness and the quality of the input from others party to the process. The NATSPEC guide does go on to provide some additional guidance but also adds that programming the specification is 'not an exact science'. We beg to differ.

It may be an obvious statement, but the time taken, and hence the effective programming of the specification process, is influenced by the manner in which the design organisation is managed, and this varies widely from the exemplary to the chaotic (Emmitt, 1999). Simply because it is a process particular to a specific design office, it is necessary for the design manager to set targets and monitor the time taken so that future programmes can be planned with more accuracy. Use of data collected on time-sheets, feedback meetings and monitoring can provide the information to allow some very accurate planning and improved quality of work. This holds true for new build and work to existing buildings.

Time and cost of producing information

Regardless of whether information is provided on paper or in digital form, both time and other resources are required to complete the task in a professional manner. Time is required to research possible solutions, to think about the consequences of design decisions, to produce and check the drawings and schedules and to co-ordinate these with each other. Time is also required for other consultants to integrate information with their own. Time will also be needed to make changes, because there will be some. Apart from all that, time is also needed to record and manage the process.

The cost of producing information is often underestimated and is not particularly well controlled in many design practices. Given the quantity of drawings that have to be produced during the detail design stage, the careful management of their production and especially of the time spent in producing them is critical to the profitability of individual jobs and will influence the long-term viability of the business. Each and every drawing, schedule and specification should be costed as a percentage of the job and allowances made for unforeseen changes, which can easily affect a job's profitability.

Outsourcing

Organisations have been quick to realise the potential cost savings and increased organisational flexibility afforded by outsourcing their non-core services. Some design offices have outsourced aspects of their work for a long time, for example specific detailing requirements to consultants with whom they have developed informal working relationships. Indeed, many design offices rely heavily on contract staff to help in busy times.

Outsourcing packages of work to other professionals can form an effective way of managing the design organisation. Some design practices are starting to specialise in design and information management, i.e. they do the conceptual design work but outsource the task of producing the project documentation to a variety of specialists, ranging from technically orientated professional design organisations to specialist subcontractors and suppliers. This is similar to the French system where detailed design work is carried out by the *bureau d'études*.

Some organisations maintain a master specification (as a control over standards) but outsource the project specification for particular jobs. Whether or not the specification writing is one of the core services provided by a design office clearly depends upon the market orientation of a particular design office. However, given the importance of the master specification, caution needs to be exercised if this service is to be outsourced because the office will effectively lose control of the knowledge and the investment contained in the master specification.

Computer-based specifications

With the growth of computer usage, increase in computer power and more sophisticated software, the potential for managing the specification process digitally has become a reality for even the smallest design office. Standard specifications, such as the NBS, and bespoke office standards are available and are widely used. These electronic formats tend to follow the same layout as their paper-based forerunners but have the advantage of being able to import information from other sources quickly from either the office master specification or manufacturers' product-specific information. Software packages offer the added benefit of being updated on a regular basis (assuming that subscription is maintained), and thus the possibility of using outdated clauses is minimised. Hardware capabilities aside, there are two essential requirements: first, ease of use for the specifier, and second, compatibility with other documentation.

1 Ease of use. Efficiency can be increased through the use of specification writing software, but only if it is simple and quick to use. Specification writing involves the transfer of information, and the easier this is to import, cut and paste within the document, the better. Search facilities are also vital to find particular words and/or clauses to help the specification writer complete the task expediently.
2 Compatibility. Because the specification is such a central document within the overall project information, it is essential that the software is compatible with other software used by the design office. It should also be compatible with software used by other participants in the design process so that information transfer and hence co-ordination can occur freely. Typical information sources are drawing files, product data library, manufacturers' technical information, current standards and codes and the bill of quantities/schedule of works. Clearly, the project specification must be based on the same software as that for the master specification.
3 Storing specification text digitally. As with a paper-based system, the files must be clearly labelled and dated to avoid any confusion. When upgrading hardware and/or software, care is needed to ensure that information (and hence organisational knowledge) can still be accessed quickly and easily.

Design reviews and co-ordination

No matter how good the members of the design team and no matter how effective the quality-control and quality-management system, discrepancies, errors and omissions do occur. Such errors are frequently related to time pressures and changes made on site without adequate thought for the consequences for other information. Many faults in buildings can be traced back to incomplete and inaccurate information and also the inability to use the information that has been provided. Discrepancies between drawings, specifications and bills of quantities can, and do, lead to conflict. Some of these can be avoided, but some slip through the net. Regardless of the sophistication of the technologies employed to minimise mistakes and ensure co-ordination, it should be remembered that people make the decisions and input the information. Thus, errors may occur.

Check-lists are a useful tool to aid co-ordination because they help to check that all the necessary information has been provided in the written specification (and highlight superfluous items). They also help to ensure that specified items are consistent with the drawings and schedules and that duplication is avoided. Offices tend to develop their own bespoke checking procedures that work for their particular way of doing things. This is particularly important when some of the production drawing packages have been outsourced because then, control at the design review becomes more critical, regardless of the quality of the outsourced work.

Accommodating design changes

In many design offices, it is common practice to write the specification as the detail design stage proceeds. It is not uncommon for design changes to be made during this period, and so these changes need to be compared with the written specification and the latter revised, if necessary. Unfortunately, some designers fail to do this, and the information received on site often contains discrepancies between details and the specification.

In an ideal world, the process of producing the production information would be a smooth affair with everyone contributing their information on time, with the information received being complete, cogent, error free and sympathetic to other contributors' aims, objectives and constraints. In reality, this is rarely the case, regardless of how good the managerial systems and the effectiveness of information co-ordination. Project teams are often assembled for one job, and the participants may not have worked together previously, and it is only towards the end of the project that teams start to communicate effectively. In the mean time, there is the potential for errors to occur simply because no empathy has been achieved. Because of this, the possibility of design changes occurring needs to be allowed for in any programming.

Of course, changes to the design can come from a variety of sources and not simply be generated internally within the design team. They can come from the client or from the contractor if the latter is involved early in the process. All changes need to be approved by the client before they are implemented, costed, their consequences fully considered, and the change recorded.

Specification substitution

An area of interest to manufacturers and specifiers alike is the problem of specification substitution or 'switch selling', the substitution of a different brand product from that originally specified. It is not unusual for a contractor to propose alternative products from those specified by brand name under traditional forms of contract. Requests for substitution on the grounds of cost should not be allowed because the contractor should have priced the job based on the original specification. Such requests may be allowed because of problems with availability or problems relating to buildability on site. In such situations, the designer (or contract administrator) has to be certain that the substitution proposed is of equal quality to that originally selected because specification decisions remain the responsibility of the design office. The original decision was often taken after an exhaustive search of similar products so that specifiers are often reluctant to approve a contractor's request because of the time it then takes to check the performance characteristics of the proposed alternative. In situations where the contract administrator is not the specifier, e.g. a project manager, the request should be referred back to the specifier for an accurate evaluation – it is unlikely that the project manager has sufficient technical knowledge to make a decision to substitute.

The problem of specification substitution has been highlighted in the technical press (e.g. Hutchinson, 1993) and by manufacturers keen to discourage such practices (Hutchinson, 1995), although, with the exception of research by the Barbour Index (1993), there is little evidence to quantify the scale of the problem. Reasons for suggesting alternative products tend to be to overcome a problem with availability, to resolve a buildability problem that has arisen unexpectedly on site, or to save money for the client (this claim needs careful consideration). Another reason is simply that the contractor is not familiar with the product and does not know where to obtain it (i.e. it is not stocked by the builders' merchant used by the contractor). As the case study (see below) has shown, it is worth checking the contractor's claims about the proposed substitution because they may not always be truthful. Less reputable contractors propose alternatives simply because they can make more money on the contract through cost savings that are not passed onto those funding the project. A particular 'trick' of contractors and sub-contractors is to wait until the last possible moment to request the change to try to put the specifier under pressure to make a quick decision. Some designers do make snap decisions over the telephone and live to regret it; others refer the contractor back to the contract clauses and their QA procedures and will not make a decision without the client's consent.

Substitution by contractors

The practice described is known as *specification substitution*. Mackinder's sample of architects had two distinct views about changing specified products when a request was made by the contractor: either they were prepared to change to an alternative product if it was of equal standard, or they refused to change the named product (Mackinder, 1980:139). Maxwell Hutchinson's (Hutchinson, 1993) advice is to stick to the original specification at all costs because the architect's office is legally responsible for any changes made to the specification regardless of who makes the suggestion (Cornes, 1983). The Barbour report (1993) found that in 56 contracts examined, just over half had experienced a change of brand name product after specification. Further evidence of specification substitution was provided in the Barbour report of 1994, in which main contractors stated that they substituted alternative products to those originally specified in 10 per cent of specifications prepared by the design team, and sub-contractors said that they, on average, altered 23 per cent of product specifications. Cost was the main reason reported for changing product specifications on the building site (Barbour Index, 1994). Apart from the Barbour reports, published research into the extent of specification substitution is elusive.

The Barbour Index (1993) also found that contractors were making changes without the knowledge of the contract administrator and that sub-contractors were making changes without the knowledge of the main contractor. Motivation for such action is financial gain, and because of this, it is difficult to obtain quantitative evidence of the true extent of such action. Another reason why people are reluctant to discuss the extent of specification substitution is that it is an act of fraud; clients are paying for specified products and getting something else. In conversation with construction managers for the purposes of this book, the extent of specification substitution would appear to be more common than suspected. It is an area that deserves the attention of some research because it can also have a serious effect on the quality and durability of the building.

In situations where the material is used externally, the planning authority quite rightly take particular interest; even so, there are many examples of contractors changing products in order to save money that have backfired because of this. In a recent example taken from a design and build project – a 12-storey office building in a prominent city centre location – the contractor changed the stone cladding to the external columns from stone to brickwork in order to save money because the contract was running over budget. This considerable cost-saving decision backfired when the planners served an enforcement notice to comply with the planning consent. That manufacturers are concerned about such actions is understandable. Manufacturers invest in product development and marketing. They also spend time on 'getting' the specification through the action of their trade representatives. Once their product is specified, they may then have to ensure that it remains so until it is built in on site – highlighted in Chapter 10. Substitution is where the manufacturers of cheaper products obtain many of their sales, and they will put considerable pressure on contractors and specifiers to get the change they desire.

Change of mind!

It is not all one-way traffic. Designers have been known to change their minds, sometimes before a job gets to the construction phase, and sometimes when the job is in progress. Care needs to be taken that the change does not repudiate the contract. The contractor will probably want some form of compensation, usually in an extension of the contract period or costs for accelerating the work to accommodate the change if it is anything other than a very minor variation. The time between the initial specification decision and commencement on site may also affect the likelihood of changes being made. In a fast track project, for example, the time span from planning permission to commencement can be short. In contrast, where the time from specification to actual assembly of the product has been lengthy, some products might be changed by manufacturers during this time, circumstances can change and costs can also change dramatically, especially in periods of over- or under-supply. In these circumstances, changes may be necessary. On rehabilitation, repair and alteration work changes will be necessitated because of the nature of the work. Changes are inevitable, but they can be minimised through extensive investigative work prior to commencement on site.

Auditing the specification process

In line with a well-implemented and managed quality assurance scheme is a need to audit the specification process both in the design office and during the contract stage to ensure compliance with the specification. In an attempt to control specification substitution, some manufacturers and researchers are looking at the possibility of bar-coding their products, for ease of identification during building and at any future date when the building may be remodelled and/or recycled and the materials recovered. Such schemes are also designed to prevent unlawful specification substitution.

Attitudes to 'new' products

Consideration of risk-avoidance and risk-reduction techniques naturally leads to a discussion about building products that are new to the specifier and/or the specifier's office. Some architect's offices have a reputation for fashionable buildings that employ the latest technological advances in materials and building components; others retain their clients by acting in a conservative fashion, using the same type of materials on the majority of their buildings. However, the architect's office is engaged to take decisions on behalf of the client, and will have to accommodate such factors as the aspirations or limitations of the client, cost control, local planning guide-lines, etc., which can limit or enhance the innovative nature of the decision. In unfamiliar situations, new or untried products may be perceived as potentially dangerous, for if they fail, there may be an insurance claim against the architectural practice. The architect as a professional is not only governed by the law of the land but also has a professional responsibility, 'duty of care', to the client. Therefore, in situations where unfamiliar products are to be selected, extra care must be taken to ensure that the product is suitable for its intended use.

Regardless of its size or type, the office is liable for its actions and is usually protected by professional indemnity insurance. Research on the specification process assumed that the perceived risk associated with using building product innovations could act as a barrier to their adoption (Emmitt, 1997). Buildings are complex artefacts and the interaction of various components and materials comes to be understood only after a relatively long period in use. The specifier will have to decide (often at the preliminary design stage) whether to employ traditional construction methods and products or risk using an innovative method. Cecil (1986) believes that a decision to adopt innovative materials, components or methods of construction presents a 'real enhancement of risk'. Not only does the specifier need to get it right, but the builder will also place greater reliance on the correctness of the specification creating additional risk. Clearly, the threat of legal action against an architectural practice should be considered, and reasonable caution should be exercised.

Theoretically, the possession of professional indemnity insurance should allow the architectural practice the freedom to take risks and use unfamiliar products, but in practice, the office has to take reasonable precautions to prevent claims being made against it. The insurance company will raise the premium if a claim is made against the office's policy and may prohibit the office from using any products that have proved defective in the past. This may prevent or discourage selection (experimentation) of unfamiliar products, despite the fact that the majority of architect's offices carry insurance cover; thus, the office *norm* may be to specify familiar products. Writing about legal liability, Hubbard (1995) used an example of a building where the roof leaked, and asked if it was the architect's fault. In liability cases, defendants have to prove that they have taken appropriate action to discharge their responsibility. Therefore, if the architect

> had specified a recognised roofing system, and if her construction details had conformed to the accepted practice, then (in fair courts) she would have discharged her obligation and should be free of fault for the leak (Hubbard, 1995:106)

The important words here are 'recognised roofing system', meaning one that other designers have specified, and 'accepted practice', meaning usual, or conservative, behaviour. There would appear to be a difference between the product that is actually new to the market and the product that has been around for some time (thus has a track record) and has been specified by other design offices. A product perceived as an innovation by a specifier may be treated differently once he or she is aware that it has been available for some time, thus reducing the perceived risk of using it. However, it is important to recognise that it is the office that holds the insurance cover, not the individual. The perception of risk to the specifier in the office may be different from (less than?) that of the partner.

Professional institutions offer advice and guidance to their members about a variety of matters, new product selection being one of them. For example, the RIBA's *Architect's Handbook of Practice Management* advises caution when considering the use of anything new, advising architects to carefully evaluate manufacturers' claims about their product's performance (RIBA, 1991).

Specifying 'new' building products

So far in the book, we have been dealing with theoretical and practical considerations relating to the specification of buildings. This chapter marks a change in emphasis. For the remainder of the book, we focus on how specifiers behave in practice, with particular emphasis on the uptake of building products that are 'new' to the recipient. This chapter starts by looking at the issue of innovation and draws on the diffusion of innovations work as a way of looking at how specifiers view products that are perceived as 'new' to them. We then define 'building product innovation' before looking at a group of specifiers' approach to specifying and attitude towards new products. The chapter concludes by reporting on research findings based on a postal survey of specifiers.

New building products

Approximately 600 patents relating to building and civil engineering are granted in the UK every year, but only 5 per cent of these patents reach production, providing the building industry with around 30 new patented products each year. In addition to these 'new' products, there are numerous minor product improvements that are constantly introduced by manufacturers to prolong a product's life in the marketplace. But getting a new product adopted is never easy. New products and product improvements are dependent upon decision-makers in the building industry for their selection, and either ignorance of the new products or conservative behaviour by specifiers tends to favour the established products.

How, then, are these products adopted? What processes are involved in the selection of a new product by a specifier? The office manager or contract administrator will need to be aware of this process if new products are to be adopted. It is also in the manufacturer's interest to understand how specifiers make their selections if they are to market their products effectively.

Early work

The focus of work on the adoption of new ideas in architecture and building design has been the province of the architectural and the economic historian, both of whom have taken a broad view of the way in which new ideas have been adopted.

Studies of medieval timber framing in England have been concerned with explaining the dissemination of the quite different regional traditions of carpentry, a process that has been seen by different scholars as both a geographical process and a social process (e.g. Mercer, 1975). While the former simply depends upon the gradual spread of knowledge of different forms among carpenters, the latter is based upon the assumption that increasing wealth allowed those lower on the social scale to aspire to the standards of their betters during a period when the timber frame was exposed. It was therefore an aesthetic consideration that drove development. The significant difference between these, but not one that seems to have been considered by the historians, is that the former implies simple technical improvements adopted by the carpenters, while the latter requires that the new forms be demanded by the customers.

Yeomans has noted that the spread of new structural forms during the late 17th and 18th centuries was influenced by the dissemination of knowledge via peer group contact, by copying and by knowledge gained from illustrated carpenters' manuals and other publications, (Yeomans, 1992:144), i.e. several mechanisms were involved – mechanisms that have also been observed by others. Peters (1988), who examined rural buildings in a small region of England, suggested that the traditions of carpentry were influenced by the spread of building books during the 18th century. Benes (1978), in a study of the diffusion of aesthetic ideas in rural New England meeting houses, although making no mention of a diffusion model, used the term 'diffusion' to advance a theory for the spread of a building style across a landscape consistent with the spatial diffusion model (Brown, 1981). Benes noted that new ideas spread into rural areas from the urban centres through knowledge gained from design books, travellers' accounts and newspapers. Furthermore, architectural fashion played a part; new meeting houses being designed to look like, or be an improvement on, those in neighbouring towns. In some cases, the same builder was employed to achieve this objective. A particular architectural fashion, and its subsequent diffusion identified by Benes, was the first use of coloured paint on the meeting houses.

The concern of the architectural historian has been to identify and explain the origins of new ideas, normally focusing on particular designers, but there has been little work that has looked at the process in general. While Yeomans (1992) was concerned with the spread of a structural idea and did make reference to the 1962 diffusion ideas of Rogers, a more recent study by the same author (Yeomans, 1996) considered the way in which theoretical studies influenced concrete mix design and made greater use of the Rogers diffusion model. In particular, this noted the influence of the leading journal in promoting the innovation to potential adopters. However, this recognised that the application of a general diffusion model to the building industry was complex because the social structure of the building industry was more complex than those involved in the studies reported by Rogers.

Innovation in this century has been important to economic historians of whom Marian Bowley has undertaken the most comprehensive studies of the building industry. Her *Innovations in Building Materials – An Economic Study* (Bowley, 1960) reviewed a wide range of products, while *The British Building Industry – Four Studies in Response and Resistance to Change* (Bowley, 1966) considered a small number of innovations in some depth and is the more significant in the context of this study. Although not diffusion studies, these provide a valuable insight into the adoption of building product innovations. Most of the innovations that she discussed in the first of these were associated with major building materials, in particular advances in their production and their associated cost benefits. She concluded that innovation in building materials is influenced by the desire of the manufacturers to hold and extend their markets rather than in response to particular needs, i.e. innovation in building products is influenced by market push rather than demand pull.

In looking at the introduction of a small product, the concrete roofing tile (Bowley, 1960), she is interested in the way in which the structure of the industry was manipulated by its producers. The Aisher family began by selling their concrete tiles as a product but found that the cost savings were being used by builders to increase their profits, rather than being passed on to customers. This failed to realise the potential sales, and in 1926, they set up as roofing sub-contractors. The success of this operation led to the establishment of the Marley Tile (Holding) Company in 1934. The implications, in the context of this study, are that an adopter may be buying into a change in the structure of the industry. Another example of a product adoption being accompanied by a change in the structure of the industry occurred with the adoption of the trussed rafter roof in Britain in the post-war period (Yeomans, 1988). In the first case, a manufacturer became a sub-contractor, and in the second case, materials suppliers became manufacturers.

Bowley's second book considers major changes, such as the introduction of reinforced concrete as a framing material or the new methods for the design of steel frames, rather than the introduction of small products. Such developments require a major restructuring of the industry or of the way in which people design. This is an interesting issue in itself but one that can only be explored as an historical phenomenon. The significance of this work is that it examines innovations that were largely rejected by the industry and considers why this should have been so. Reinforced concrete failed to achieve the market share that it might have against steel frames that were established first, and engineers resisted changes in design methods proposed by the Steel Structures Research Committee.

Other economists, such as Stone (1966), have looked at the relationship between innovation and the cost of labour and materials and found that innovation in building has been generated in different ways: clients have set new problems, designers have used new materials to solve problems and contractors have used new materials to reduce the cost of construction. All these have led to an increased range of materials and increased the number of possible methods of building, suggesting a demand pull: apparently contradicting the views of Bowley.

The modification of existing buildings has also been an area for the adoption of innovations. In Britain, the adoption of plastic-based window frames was associated with the retrofitting of double glazing in houses. The adoption of solar heating by households in a California neighbourhood was studied by Rogers and used as one of his 'case illustrations' in the third edition of *Diffusion of Innovations* (discussed below) to highlight the influence of networks on the diffusion process (Rogers, 1983:300–304). Although it dealt with building products, the study was essentially based on household consumer behaviour, as the solar panels were installed into the roof of existing houses. In this example, it was the visibility of the product that affected its rate of adoption.

The maintenance and upgrading of buildings, of which the above are just two examples, are operations that encompass both professional activities, where corporate clients are involved, and the ordinary consumer who may be buying direct from the manufacturer or working with the assistance of a builder. These may involve very large volumes of sales and so be significant developments in the industry. In the professional field, it is flooring and partitioning that are the major product types involved in retrofitting. In recent decades in the UK, double glazing became a large seller in the ordinary consumer market. While these kinds of products are innovations, the upgrading of commercial properties is often carried out by a quite different group of professionals from those involved in new buildings so that this and the domestic consumer market lie outside the scope of this study.

Few of these studies have considered the behaviour of the designer of buildings, i.e. the specifier of new building products, in any detail. The starting point for historians who are looking at the new ideas of particular individuals has been the act of adoption, attempting to explore the possible origins of each idea. The implicit assumption is that what is of interest is the process of transmission of ideas that are actually used. But one might equally ask why ideas are not adopted. If an idea already exists for a long time before it is more generally adopted, then an understanding of the adoption process also needs to consider why it was rejected by those who were previously aware of it. As Yeomans (1992) has shown, a problem for the adoption of technical innovations is the conservative behaviour of apprentice-trained craftsmen who are likely to cling to what they already know rather than to adopt the unfamiliar. They may only change when forced to do so as a result of external pressures. Both his recent study of concrete design and Marian Bowley's study of developments in frame structures showed that professionals have behaved little better. If we are to understand these adoption processes, then our attention must be focused on the behaviour of these professionals.

Today, few buildings use only the simple, basic materials like bricks and timber. The vast majority use a wide range of manufactured products. Even basic components like the screws and nails used to fasten other components together have undergone improvements that make them quite different from their predecessors. The principal source of information for the designer on these new products is the extensive body of descriptive material produced by building product manufacturers, i.e. their trade literature, which is still the basic starting point even though electronic forms of information are becoming available. Therefore, the natural starting point for any enquiry is the relationship between this information source and the behaviour of the specifier. The first questions concern how this material is made available to the specifier, i.e. the nature of the communication channels between manufacturers and designers. How readily is material available during the design process itself, and what is the behaviour of the designer in using this information? It is with the availability and use of the material in the design office that we need to begin.

Well-managed offices will commonly have a library of product information; a central resource that the members of the office can and are expected to use. However, experience shows that many designers will have their own collection of product information, even though this may be in contravention of office policy. In spite of the availability of this resource, it would be wrong to assume that it is part of a designer's normal behaviour to search the product literature in order to find just the right product for the particular task that he or she has in mind; instead, behaviour is rather more conservative. Moreover, office libraries and personal collections are themselves highly selective.

Because designers tend to use familiar products, it may be some time before they even become aware of more recently developed alternatives. If this conservative behaviour results in the formation of a personal palette, how does a building product that is perceived as new by a specifier working in a designer's office get to be specified in preference to one already familiar? Apart from Mackinder's (1980) work, there has been little research into the process by which specifiers become aware of products with which they are unfamiliar or into the way in which these products may be adopted in the decision-making process that follows. It is necessary to look outside building to fields where the selection of products has been studied to obtain a better idea of this process. For example, to examine how a product is adopted, a natural starting point would be to look at marketing literature, in particular that concerned with consumer behaviour (e.g. Chisnell, 1995). The problem with this is that it is concerned with 'new' products, i.e. those recently launched onto the market. While specifiers may consider building products that are new to the market, they may also consider products that have been available for some time, but which they have only just become aware of. This can occur when a specifier is faced with a new kind of problem and needs to use a range of products of which he or she has no previous experience, simply because they were not required for previous jobs. Then, even those products that other specifiers have experience of might then be new to this particular person.

It is the body of work on diffusion of innovations (Rogers, 1995) that is concerned with an individual's reaction to new ideas and examines the mechanisms of adoption or rejection. This work also treats an innovation as something that is perceived as new, whether or not it is in fact new. It is the newness of the idea to the recipient, rather than the length of time that it has been on the market, that sets diffusion literature apart from marketing literature. Because diffusion theory is concerned with the factors that influence the rate of adoption of ideas or products that are perceived as new by the receiver of the information, the potential adopter (in this case, a specifier working in a designer's office), it provides a general model that is more relevant to the behaviour of the designer/specifier.

Diffusion research

The spread of new ideas, practices and products within a social system is known as the diffusion of innovations, and the large body of literature concerned with this (Rogers, 1995) is discussed in greater detail later. In simple terms, diffusion studies are concerned with the communication of an innovation, to a social system over time, described by Rogers (1995:5) as:

> ... the process by which an innovation is communicated through certain channels over time among the members of a social system. It is a special type of communication, in that the messages are concerned with new ideas.

At the heart of all diffusion research is the adoption process, in which an individual or group of individuals become aware of an innovation and react to it, either by adopting or rejecting it. Reaction to the innovation is not an instantaneous act, but a process that continues over a period of time with distinct stages known as the innovation-decision process (Rogers, 1995:161–203) (see Figure 7.1). It is the cumulative effect of adoption of an innovation over time that results in the classic diffusion curve.

Awareness of the innovation will come from one of two communication channels, either mass-media channels or interpersonal channels (Rogers, 1995:17). Empirical evidence has shown that the former are more effective in raising awareness of innovations, whilst the latter are more effective in changing attitudes and actually influenc-

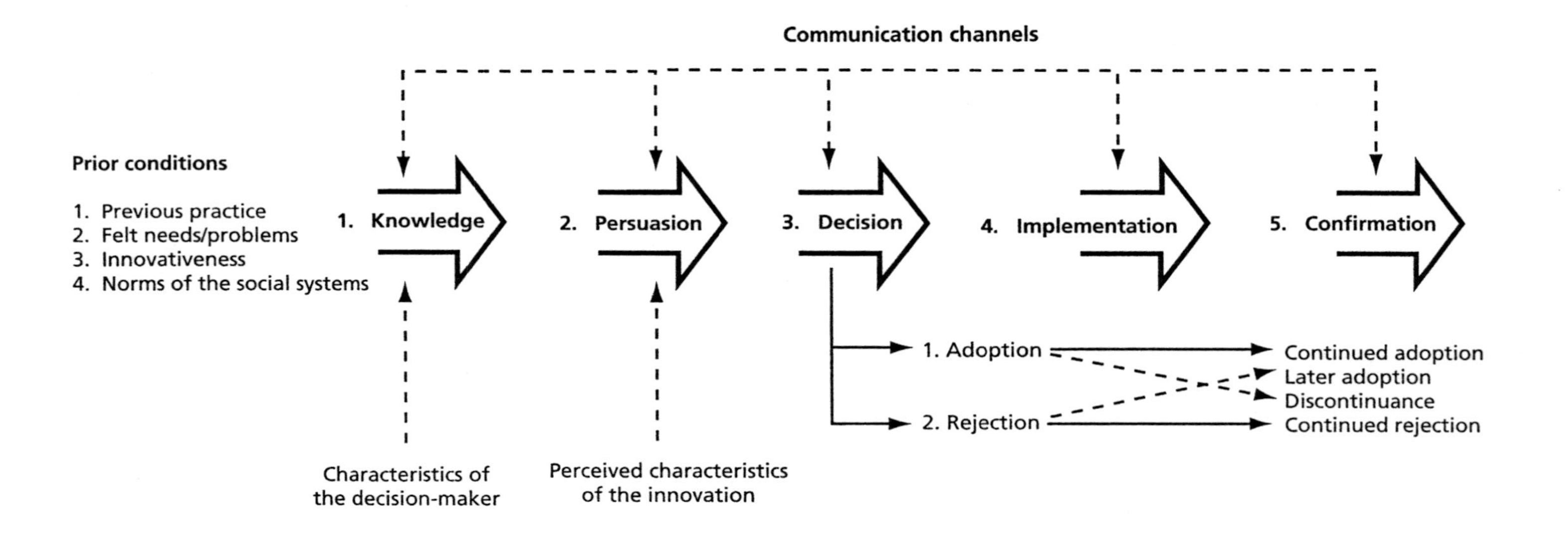

Fig. 7.1 The generation and diffusion of innovations (based on Rogers, 1995 : 163)

ing the decision to adopt or reject the innovation. Rogers has shown that the innovation will have a number of perceived characteristics, which will influence its rate of adoption, (Rogers, 1995:15–16) but also that the individuals in the social system who are exposed to knowledge about the innovation will themselves have different characteristics. Based on empirical findings, the individuals in a social system have been classified into five categories, ranging from the first to adopt, the innovators, to the last, the laggards, according to their degree of innovativeness (Rogers, 1995:22).

Since specifiers must be aware of a building product in order to specify it, the communication of information about building product innovations to the potential specifier is critical to their adoption. An inquiry was therefore conducted into the communication of building product innovations to the architect's office and attempts to measure the specifiers' reaction to these products identifying the factors that result in their adoption or rejection. This is a part of the design process that has been largely overlooked, or ignored, by the design methods authors, the detail design process, during which building products are usually specified (discussed in Chapter 2). The intention of the inquiry was to use the E. M. Rogers diffusion of innovations model (Rogers, 1995) as a structure for the work, the objective being to try to establish how specifiers respond to building products that are new to them. It did not attempt to address the level of innovativeness of architects, nor was it directly concerned with how the diffusion of innovations affect the design, function or appearance of the building. It was an attempt to understand the behaviour of architects or other specifiers (acting on behalf of their clients) of building products that are new to them, i.e. it was primarily concerned with the innovation-decision process.

However, it became evident that a simple application of the Rogers model was not possible because of the special nature of the building industry; in particular, this industry has a rather more complex social structure than those previously studied and reported by Rogers. As a result, the focus of the research was adjusted to re-consider the applicability of the model to the building industry. Therefore, a theoretical extension of the Rogers (1995) diffusion paradigm was developed, supported by a number of data collection exercises, both for illustrative purposes and as a means of testing the theoretical model. In the detailed examination of the main principles, the following questions were asked:

1 How is information concerning building product innovations communicated to, and received by, the architect's office, i.e. how does a specifier become aware of an innovation?
2 What are the specific characteristics of the social system being studied, and how do they influence the diffusion of building-product innovations?
3 To what extent is the specifier's innovation-decision-making process influenced by factors other than his or her personal preferences?

4 How do building products that are perceived as new to a specifier get specified if he/she relies on the palette of favourite products?

These questions were addressed largely through the use of the case studies reported below, which were supported by quantitative data collection. From these, a theoretical model has been proposed, which is based upon the Rogers model, modifying this to suit the particular structure of the building industry.

The generation of innovations

For a product to be diffused, it must first be developed, manufactured and launched onto the market. This subject area has been covered extensively, from invention (e.g. Gilfillan, 1935), through product development (e.g. Bradbury, 1989) to marketing (e.g. Midgley, 1977; Druker, 1985), and a new idea or new process adopted by manufacturing industry has been described as a technological innovation (Utterback, 1994) or a process innovation (Davies, 1979). These innovations are concerned with the introduction of new machinery or production methods and their effect on productivity, and because of this, they tend to be studied by economists such as Bowley (1960). Studies concerned with the manufacture of building materials have been carried out by Davies (1979), who included a study of the brick-making industry in his work, while Layton (1972:80–93) investigated the introduction of the float-glass process by Pilkingtons.

Parker (1978) referred to the development of new products as the innovation process, and divided the process into four aspects: invention, entrepreneurship, investment, and development. Other authors, such as Bradbury (1989), have made a distinction between the initial idea (invention), and the innovation process, which covers all stages of a product's development up to, and including, its launch onto the market. Although this was a process effectively ignored by Rogers until the third edition of *Diffusion of Innovations* (1983), it has since become an essential element of the diffusion model, the generation of innovations (Rogers, 1995:131–160), which ends in a decision by the manufacturer to market the product to potential adopters. It is the decision to market the product that is the start of the diffusion process (see Figure 7.2).

Apart from innovation that has come entirely from manufacturers, there is a long history of architects' involvement with these initial phases of the process. Holden (1998) has shown how designers of Lancashire cotton mills in the 19th century were concerned with the development of fire-proof flooring, while Saint (1987) describes how, a century later, those involved in the post-war school-building programme involved themselves directly with manufacturers in the the design of flooring, sanitary ware and furniture for their buildings. Indeed, there may be a complex relationship between the development of architectural ideas and that of suitable products through

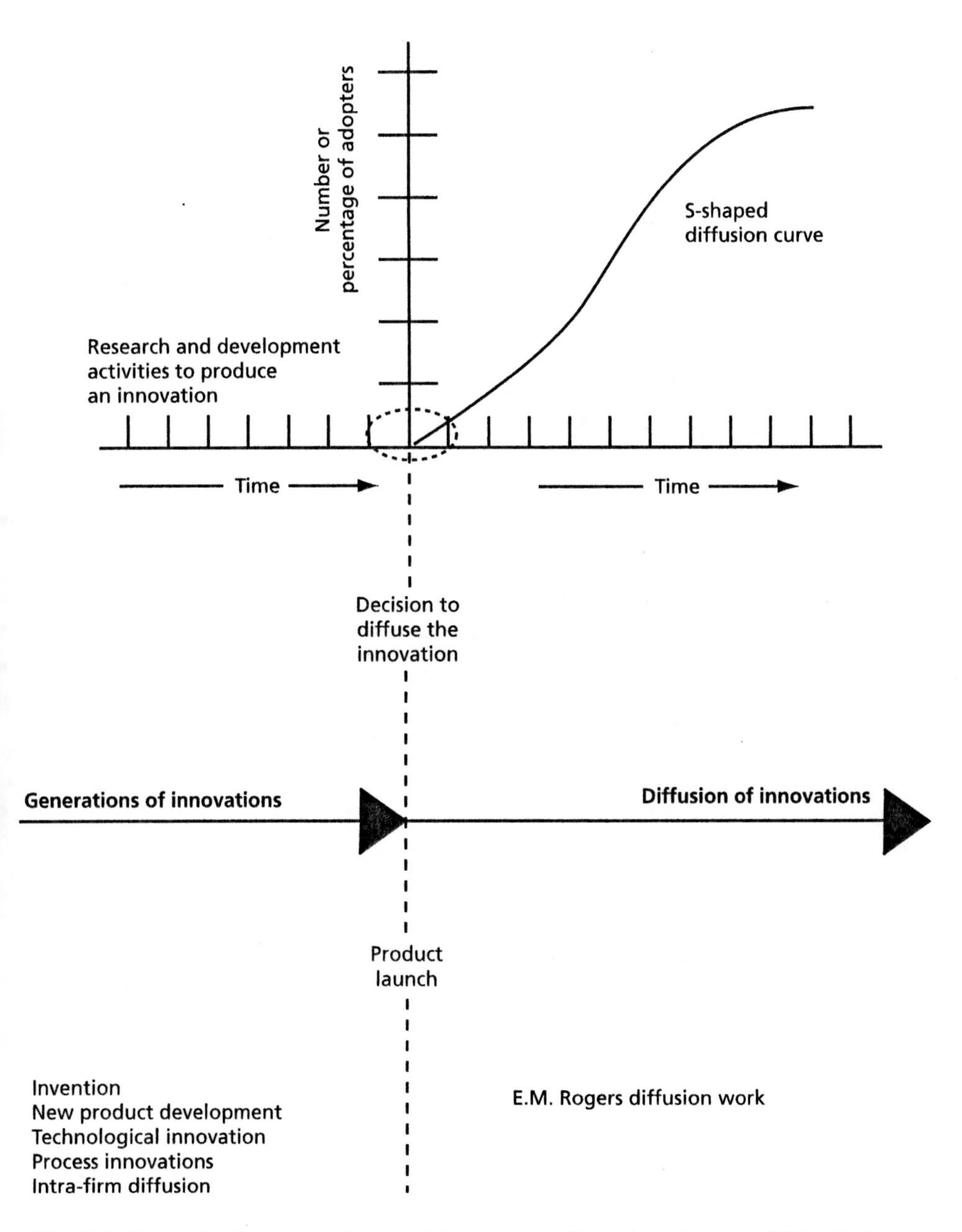

Fig. 7.2 Stages in the innovation-decision process (based on Rogers, 1995 : 110)

which these ideas may be realised. The initial ideas in the development of cladding systems were undoubtedly architectural, but they could not have been realised in the way in which they were without the active involvement of manufacturers of cladding components. What is not clear is the extent to which architectural ideas have played a part in the development of other building products. As an example, we might cite the development of devices for fixing brickwork to the structure of buildings as architects in the 1970s began to treat this material in a far more plastic manner. Here, too, was a prima-facie example of architectural demand driving the development of building products.

There is a chicken-and-egg issue here associated with visibility. Products like those used to achieve suspended, non-structural brickwork shapes that derive from a particular architectural fashion are the kind of items that can be produced only as a result of demand pull because they depend upon the development of a particular architectural fashion. Here, the innovation is an architectural fashion, but one whose adoption is also dependent upon the availability and use of innovative building products with all the attendant risks both to the specifier and to the manufacturer. The issue of visibility is that while the product may not itself be visible, it has visible effects. If brickwork is used in a sculptural way like concrete, then, to any architect, there is an intimate connection between the visible form and the kind of products that, though invisible, must have been used to produce that form.

There is another possible process to consider here, which is the extent to which architects are able to persuade manufacturers to produce the things that they want. Banham (1969, 204) cited private correspondence from J. R. Davidson, who claimed that in the late 1920s, it was difficult for architects to persuade manufacturers to produce the kind of light fittings that they wanted. More recently, Oostra (1999) has presented a case study of work carried out in Holland to develop window mullions using a new kind of material – new, that is, for window mullions but widely used in the manufacture of sports goods. In spite of the successful development of these components for a particular project, the idea could not be developed more widely because of lack of interest on the part of the manufacturer. The manufacturer who used the material in the production of sports goods seemed unwilling to enter the building industry, perhaps on the principle of 'cobblers and lasts'. The difficulty is that one is always dependent upon such anecdotal evidence, and we have insufficient well-researched case studies to know what kinds of conditions favour the uptake of architects' ideas by manufacturers.

The significance of this case study is simply that it shows that there are architects who are not overtly adverse to innovations. Quite the contrary, as architects see themselves as innovative, one might expect them to be willing to embrace the new ideas of others. Nevertheless, Bowley's (1966) studies have given us an image of an industry that is resistant to change. It is a matter of commercial prudence that produces this conservative behaviour. At the same time, one must assume that manufacturers are conscious of the need for their products to be acceptable to architects and therefore to be responsive to their comments. If this is a significant aspect of a manufacturer's marketing methods, then it should only be a small step from the modification of products to take account of feedback from architect to the development of new products to meet their needs. However, the extent of this as an important generator or modifier of new building products has also not been studied and is beyond the scope of this investigation.

Definition of 'building-product innovation'

So far, the word 'innovation' has been used rather loosely. *The Concise Oxford Dictionary* (1990:610) describes the verb 'innovate' as 'bring in new methods, ideas, etc.; make changes' and 'innovation' and 'innovator' as 'make new, alter'. The synonyms listed against 'innovation' in the *The Oxford Thesaurus* (1991:223) are 'novelty; invention; modernisation; alteration'. The word innovation is used in different ways by different authors to mean different things. Even within a single industry or single profession, innovations can be of many different kinds, concerned with new ideas, new products or new methods. Furthermore, authors concerned with different subject areas, such as economics, politics, sociology, design, engineering, corporate management, marketing and consumer behaviour, all use the word differently. In architectural literature, the word innovation tends to be used to describe either the design approach of the architect or the appearance of the finished building. For example, architectural journalists often refer to the design of the building as 'innovative' or state that the architect has worked in a manner regarded as 'innovative' by his or her peers. In architectural literature, therefore, the word is often used as a substitute for 'creative' and does not have the same meaning as the word in diffusion literature.

Marian Bowley (1960), divided innovations into two main groups, 'those that change the product and those that affect costs and availabilities' (1960:25), and was concerned with innovation as viewed by the consumer, the building user. She classified innovations in a range, from those that result in new products to the consumer (not substitutes for existing products) to innovations that led to products that were, from the viewpoint of the consumer, no different from existing products (a perfect substitute). Bowley went to great lengths to classify innovation (pages 25–43 of her study) concluding with the observation that '... there are innumerable ways of working out classifications of innovations, and the advantage of one rather than another depends on the particular purposes of the study.' (1960:43). Bowley's sub-divisions (1960:25–43) are also different from the way that the term innovation is used in diffusion literature. Since the intention is to apply the diffusion model, the classification used here will follow that used by diffusion researchers.

Rogers is concerned with the total population of a social system, i.e. all potential adopters, and has defined innovation as a product or idea that is new to the recipient, regardless of how long it has been available. Since the research is concerned with the perception of specifiers working in architectural offices, for the purpose of this research, the definition used by Rogers (1995:11) can be re-written as:

> An innovation is a building product that is perceived as new by a specifier. Whether or not the product has been recently launched onto the market is not important, it is the perceived newness of the product by the specifier that determines his or her initial reaction to it.

The work of Everett M. Rogers

If the economic process depends upon the capacity of industry to continuously develop new products and new processes (Druker, 1985), it follows that it is equally dependent upon their adoption by the consumer, and it is this process that has been explored in diffusion research. The history of diffusion research is well documented (Rogers, 1995:38–95), the main subject areas covered being anthropology, early sociology, rural sociology, education, public health, medical sociology, communication, marketing, geography and general sociology. Published in their own field, these diffusion studies were concerned with the uptake of both innovative ideas and innovative products.

Gabriel Tarde (1903) was recognised as the first to investigate the adoption or rejection of innovations. His publication *The Laws of Imitation* identified several of the main issues of diffusion, from the S-shaped curve to the important role of the opinion leader in a social system. While the influence of Tarde is still present in diffusion studies, the start of diffusion research dates back to the 1930s, with the majority of early work undertaken by rural sociologists. Given this tradition, it is not surprising, therefore, that a rural sociologist was the first to publish a comprehensive book on the subject in 1962 with *Diffusion of Innovations*, since which time, Rogers has continued to publish in the field.

The first edition of *Diffusion of Innovations* (Rogers, 1962), which summarises diffusion literature, has been described as a 'benchmark study' by fellow diffusion researchers (e.g. Brown, 1981). It consolidated the work of 405 separate publications, including 27 of his own, provided generalisations and definitions that could be used universally and provided a single model of the diffusion process that others could, and did, use. Rogers continued to update and revise his work as the number of individual diffusion studies increased and a second edition, *Communication of Innovations – A Cross-cultural Approach* (Rogers and Shoemaker, 1971) was co-authored with Floyd F. Shoemaker and had a strong emphasis on the communication process, reflected in both the revised title and the influence of Shoemaker. The third edition marked a return to the single author and original title. By the time the fourth edition of *Diffusion of Innovations* was published in November 1995, there was a total of 3810 independent publications.

The main principles of diffusion work categorised by Rogers (1962, 1983, 1995) and Rogers and Shoemaker (1971) have been used by the majority of subsequent diffusion researchers whose work has ranged from simple models to sophisticated models based on complex mathematical formulae. For example, in geography, Brown (1981) uses the work of Hagerstrand (1969) and Rogers to develop his own paradigm of diffusion across the landscape; spatial diffusion. Others have either applied the model to different fields such as manufacturing and product development, or have concentrated on a small part of the process. For example, Foxall (1994) has concentrated on the characteristics of adopters, whilst Gatignon and Robertson (1991) have concentrated on inter-personal communication in the development of their consumer diffusion paradigm.

However, there has been virtually no work on the building industry, and therefore, it may be helpful to look for parallels in other areas, that is situations that may have some similarities with the building industry. There are two distinct areas within diffusion literature that might be considered to have some relevance: research into the diffusion of new products (e.g. Bass, 1969; Mahajan and Wind, 1986) and research into the adoption of drugs by medical doctors (e.g. Coleman, 1966). Although work on the diffusion of new products has drawn on the Rogers model, it is specifically concerned with products that have been launched onto the market recently: and given the special nature of design work (discussed above), it is unfortunately not relevant to this work. It is the research into the adoption of drugs by medical doctors (Coleman, 1966) that is more relevant because it is work that has been carried out into the adoption behaviour of a professional group that has parallels with the architectural profession. Doctors have a duty of care when selecting drugs on behalf of their patients just as architects have a duty of care towards their clients when selecting building products.

Early clues and questions

At the beginning of this research a 'focus group', or group discussion, which one of the authors attended, was arranged by a commercial market research firm. Their aim was to investigate the reading habits and behaviour of architects with a view to restructuring the editorial content of a long-established weekly publication with a strong practical bias in its content. It regularly reviews building products and reports on the technical aspects of the buildings that it describes as well as carrying a lot of advertising space for manufacturers; therefore, most of the discussion was directed towards the specification of building products. The nine architects who took part in the group discussion comprised two partners/directors in medium to large architectural practices, five associates in small architectural practices, and two solo practitioners. All of the offices were located in the same geographical area, but none of the architects knew one another personally. Although a very small sample, this suggested that architects' offices are not linked by a network so that the possibility that knowledge of building-product innovations might be diffused by social interaction between offices is unlikely.

Apart from the two solo practitioners (who carried out all aspects of the architect's duties), the architects said that they were not involved in building-product selection themselves, but were responsible for supervising or overseeing the work of less senior members of their office who were. Product specification was seen by them as the least glamorous and most tedious job in the office, as well as being a particularly time-consuming task, and tended to be delegated to the lower-paid members of the office, thus confirming Mackinder's observations.

When questioned on the subject of specification, all stated that the office selected familiar products. This was reinforced by the use of standard specifications that were applied to all jobs and altered to suit the specific criteria of a particular project. These specifications had been developed over a number of years and contained what each practice considered to be the best materials or products for each purpose, based on previous experience. Products included in standard specifications had been specified and built into previous projects and so were both familiar to the office and known to perform the function for which they were specified. All of the sample said that they preferred to specify products by proprietary name rather than by performance specification and were very reluctant to change products once specified. This was because of the time required to assess any new product.

All said that a product that was new to the office would only be specified if it was supported by guarantees, British Board of Agrément Certificates, and/or conformity to British Standards. Manufacturer's guarantees were seen to be essential in reducing (perceived) risk, thus minimising the possibility of an insurance claim for specifying a defective or unsuitable product. Furthermore, they said that they would only use new building products if absolutely necessary, preferring to wait until someone else had used it. Products that had been specified on past building projects and had not performed as stated or had failed were deleted from the standard specifications and effectively blacklisted.

The participants said that products could not be compared on a cost basis because the cost of a product was rarely known at specification stage. This was because manufacturers were reluctant to disclose the cost of their products until after they had been specified. Feedback on overall project costs was provided by the quantity surveyor and, sometimes, the contractor when he wanted to change a specified product for a cheaper alternative. This allowed the architect to build up an elementary knowledge of costs. The two solo practitioners, who carried out work for small-scale developers and builders, claimed to have a better knowledge of costs because they had to work without the services of a quantity surveyor. They claimed that they rarely used products that were new to them because of the time required to investigate them fully and the threat of legal action should they fail. Neither of the solo practitioners carried professional indemnity insurance and said that they therefore had to be particularly careful in their selection [1].

The primary source of awareness of new products was the office library, which, for larger offices, was administered on a part-time basis by a servicing agency, with varying degrees of input from other members of the office. This helped to ensure that the library was updated, usually on a monthly basis, with manufacturers' up-to-date literature. After the library, the trade journals were the next source of reference. The journals were generally used when looking for sources of inspiration for the design of a particular building type, with less emphasis on product selection. They all said that they rarely took any notice of advertisements in the journals but did say that they would occasionally send off for more information using a 'reader reply card'. Trade fairs and exhibitions were rarely attended because they were seen as a waste of time.

The architects were then asked about the sources they used to select products. Past experience of the architect and of the specifiers in the office was the main source of information for all of the sample. Six said that they occasionally used trade literature in the library, whilst one sometimes followed up articles on projects featured in the journals. In a general discussion, prompted by the organiser, the sample said that occasionally, they would use a product recommended by a client or a consultant. Trade representatives would only be invited into the office once a genuine interest in the product had been established and only where it was relevant to a current job. If the representative called speculatively, they were usually asked to leave their literature but rarely seen by specifiers in the office because of the lack of time available. Generally, the architects were dismissive of the representative, and it was this experience that eventually led to the gatekeeping research (reported in the next chapter), which confirmed this anecdotal evidence.

The postal questionnaire

It was evident at an early stage that there was little published material that related directly to this subject, and the simplest way to begin was to gather some preliminary information using a postal questionnaire (see Appendix). The questionnaires were addressed to senior architects and partners of the architectural practices.

The purpose of the postal questionnaire was to test ideas that were being developed in the theoretical model. A traditional form of contractual arrangement was assumed

1. At the time, professional indemnity insurance for architects was not a requirement of the professional body.

in this, and that specifiers selected building products by brand name rather than by generic terms. Therefore, it was important to establish the extent to which these assumptions were borne out in practice. Answers to three questions supported a general model based both on the use of traditional contracts and on product specification by brand name.

Over the previous 5 years, traditional contracts had been used more than any other contractual arrangement; 92 per cent of respondents indicated that it was their first choice, with design & build liked by only 8 per cent. It had also been assumed that the type of contract would influence the manner in which building products were selected, and question 9 was designed to see whether respondents believed the same. Twenty-eight per cent of the respondents, when asked if they thought that the type of contract influenced the selection of building products, answered 'no', while a further 29 per cent answered 'generally not'. Of course, given the preponderance of traditional contracts, they may not have had sufficient experience of other types to know whether selection of products would have been different or not. However, the same high percentage of traditional contracts among their recently completed jobs meant that their experience of selection procedures would not in fact have been particularly influenced by contract method.

There was a strong preference for the use of precise trade names (brand names), with 85 per cent always or often selecting materials by this method. However, 58 per cent claimed that they always or often selected materials by a generic description. Thus, some respondents said that they selected by both methods, and this suggests that the two methods are both used concurrently for different product types.

An attempt to record the influence of office policy on product selection was made, and the presence of an approved list of products was noted in 31 per cent of the practices, whilst slightly more offices (34 per cent) operated a blacklist of products. Approved lists or blacklists of materials and products were not as widespread as envisaged, but it was noticed during the group discussion (reported above) that architects were reluctant to admit to the existence of such blacklists, and it would be difficult to devise a means of testing for their presence. Therefore, this response should be treated with caution.

Stage 1. Knowledge

Of the journals read, *Building Design* and the *Architects' Journal* are the most popular, both of which carry a limited amount of product advertisements and articles referring to products. *What's New in Building* and *Building Products*, both issued free of charge to specifiers, were popular; known as 'product journals', they primarily carry advertisements and articles relating to building products. Journals were thought to be the most important source for supplying details of new building techniques and products/materials, followed by direct mail, trade representatives, office library and colleagues. However, as we shall see, this belief is not necessarily borne out in fact.

There is a problem here because advertisements in journals do not give many details. Certainly, specifiers receive information about new products by this means, but they will not obtain any technical details unless the journal makes a policy of describing products in detail. Of course, the *Architects' Journal* in Britain does just this with its *Products in Focus* series, but this is a special case and subject to editorial policy. British specifiers are also served by the free publications that specialise in providing product information. This means that the significance of journal information may be higher here than in countries where available publications have a different editorial policy.

The majority of the respondents said that trade journals influenced their design decisions and their selection of materials, although only 35 per cent thought that they were likely to select a material or product on the strength of an advertisement or a technical article in a journal. However, unprompted comments such as 'No, but instigate checking on it' and 'No, further research needed and Agrément certificate to be examined' reinforced comments from the group discussion that awareness was raised by articles and advertisements contained within the journals.

Stage 2. Persuasion

A high percentage of the respondents (92 per cent) claimed that they consulted trade literature on a regular basis, but more significantly, 80 per cent confirmed that they kept their own file of product information. This is important, because the presence of a palette of favoured products could be a barrier to the awareness and subsequent adoption of product innovations, especially those marketed by a company not included within this personal library. The question that asked if they would expect to be able to specify a product on the basis of the trade literature was an attempt to establish architects' expectation of this source of information. In fact, they were surprisingly high, 63 per cent expecting to be able to make a full and detailed specification on the strength of product information. Unprompted comments reiterated their expectations, from 'Yes, vital', suggesting that if it was not possible, he may look elsewhere, to 'Expect yes, more often though it cannot be done.'

This suggests a pattern of behaviour rather than an expectation about the nature of trade literature. What specifiers seek is the ability to carry out this part of their job as simply and quickly as possible. They will be aware of the additional time needed to seek out information on the product from other sources, and so the quality of the product information is in part part of the selection process. What they actually 'expect' in another sense of the word may be quite different. The comments on the actual quality of trade literature suggest that in that sense, their expectations based upon experience are quite different.

The majority of the respondents said that they call a trade representative to assist with the specification: 57 per cent sometimes and 25 per cent always. This underlined the importance of the trade representative as an agent of reinforcement. Again, two unprompted comments are of interest, 'Yes, often have to, unfortunately; literature often inadequate', reflecting the observations made above, and 'Representative requested to assist if product unfamiliar …', which suggests that the trade representative is more likely to be contacted when there is a high level of uncertainty about the product.

Loyalty of the specifier to a favourite manufacturer was tested by asking whether, in circumstances where a manufacturer normally used does not produce the exact requirement, the architect attempted to find an alternative, compromise and specify the familiar manufacturer, or ask the manufacturer to revise the product. Although some indicated two answers, the majority, 74 per cent, would attempt to find an alternative manufacturer, that is, brand loyalty is not particularly strong amongst specifiers: only 12 per cent said that they would compromise and stay with the familiar manufacturer.

Stage 3. Decision

There is a question of whether specific materials or components are requested by people external to the architect's office. The respondents said that the client was the most active in this (26 per cent recorded 'often' and 54 per cent 'sometimes'). Planners were the next most frequently responsible for requesting particular materials or components, closely followed by the contractor. According to the sample, the quantity surveyor was not regarded as having much influence, although this contradicts Mackinder's findings.

An aspect of the responses to this question that is not clear is when such outside influences occur. The client may require certain materials to be used as part of the brief. Also, the planning officer may have an early influence. However, requests from quantity surveyors and contractors may be late in the process, made in an attempt to reduce costs.

Awareness of costs was shown to increase as the project progressed from scheme design through to detail design/specification. This is in line with expectations. Just under half of the sample (44 per cent) said that they were aware of price differentials between similar products, and a further 44 per cent said that they were occasionally aware of price differences, but we cannot be sure how early this might be in the process. When asked if product cost would influence final selection, if they were aware of a range of product costs, 56 per cent said that it would often and 36 per cent occasionally. However, there is evidence of a reluctance on behalf of some manufacturers to disclose prices to the specifier as indicated by the unprompted response that 'Suppliers and manufacturers (are) often reluctant to make current prices available.' Another unprompted comment was that 'No two products or subcontracts are ever the same', suggesting that it is difficult to compare like with like. These observations are in line with the experience of one of the authors. It is often extremely difficult to obtain prices from manufacturers, and those quoted to an architect may differ from those given to a contractor. While many architects may be aware of prices, it is not clear from the survey exactly how this information is obtained. In some cases, it may be from previous jobs.

Over half of the respondents said that they had changed a material, product or component on their last project as a direct result of a request by the contractor. This in itself is high, demonstrating that the specification process continues whilst the job is on site and shows that changes may be made under pressure. The most common reason given for changing was non-availability and/or unacceptable delivery times (to the contractor), although what evidence the architect may have for this is uncertain. While such changes have ramifications regarding the architectural office's liability, it also highlights the contractor's contribution to the process. Contractors appear to exploit time pressures as a way of influencing or forcing a decision that might otherwise not have been considered. Thus, the contractor can form a barrier to building product innovations by leading to their discontinuance. Alternatively, he can introduce knowledge of a product innovation to the specifier.

Other reasons for changing products during the contract at the request of the contractor were because of cost and to suit the contractor's programme. There were also a few examples of change due to construction difficulties on site. Some comfort to those attempting to introduce new products through architects may be drawn from the respondents who confirmed that they had not deviated from the tendered scheme, best summed up by 'Change nothing if possible', reflecting the fear of liability, additional work and potential problems.

Product characteristics

An important element of diffusion research is the length of time for a particular innovation to diffuse within a social system. Caution was emphasised in the answer to Question 18, with the majority waiting over 2 years for the product to be on the market before adoption. Prompted comments ranged from 'Depends upon type of product, a brick has little to prove', to 'The longer the better – particularly if completely innovative', thus confirming the conservative behaviour reported by Mackinder.

Caution was noted when selecting products that were previously unknown to the specifier; 49 per cent occasionally select products of which they had no previous experience. But this depends to an unknown extent on the type of product summed up by the comment 'wall ties no, wallpaper yes'. The majority of respondents indicated that they would wait until someone else had specified a product before specifying it themselves, indicating that peer group approval is an important factor in the adoption of innovations. The responses also indicate that risk avoidance is an important consideration.

Only a little over a third of the sample had specified products that were new to them or new to the market within the past year. The list of materials recorded indicated that a new product to one specifier may not be to another, and furthermore, some of the products recorded as new to respondents have been available for a considerable period of time. There was no attempt to classify the respondents in terms of their speed to adopt innovations. However, this is an issue addressed in the diary of adoption.

Only 31 per cent confessed to specifying products that were completely new to them, but an interesting list of products, components and materials was noted. Interestingly, one answer to this question that was indicative of the earlier caution was, 'Yes, too often', suggesting that problems had occurred. This was followed by a general comment; 'Use of innovative products is often restricted, not by any doubt on performance, but by insurance companies indirectly and contractors unfamiliarity directly'.

There is an interesting speculation possible here. Because control in France is exercised through the need to insure buildings and because the insurance companies, rather than the local authority, check the designs, there may be a greater resistance to product innovations – perhaps a greater reliance on Agrément certificates.

The need for observational methods

The postal questionnaire was useful in identifying and quantifying some of the issues, but it was clear that some naturalistic methods of enquiry were required to look at the specification process in more detail. These are described in Chapters 8–10.

Becoming aware of new products

New products are dependent on specifiers for their uptake and commercial success. Because specifiers need to be aware of products before they can consider selecting them, the efficiency of the manufacturer's marketing campaign is important. The variety of tools used to communicate information to potential specifiers is reviewed, as is the design office's need to manage the information coming into the office through the use of technological gatekeeping. The chapter concludes by looking at specifiers' apparent resistance to information about new products and the implications of such behaviour.

The marketing campaign

Manufacturers are in business to sell products and to make a profit. In order to do this successfully, they must identify a market demand for their products and then exploit this by utilising the communication channels available to them, an activity generally referred to as marketing. Once a manufacturer has made a commercial decision to market a new building product, the primary task is to create both interest and subsequent demand from potential adopters. But, as we have seen, products are marketed to, and specified by, different members of the building industry. For example, some products will be designed to be sold only through builders' merchants, whilst others may be designed to be sold only to architects, or a combination of both. Therefore, the marketing of a product, will largely determine those likely to become aware of it, which in turn will influence who is likely to adopt. So far, we have discussed processes that occur largely within the design office, the filtering of information, the search for information and the contacts with manufacturers that are largely initiated by the designer. What we have not yet considered is the nature of the advertising material generated by manufacturers as part of their marketing strategy and the extent to which specifiers are conscious of this material.

Manufacturers of building products rely, primarily, on trade literature and trade representatives to raise the awareness of potential specifiers to their products. Trade literature covers a wide range of information produced by the building product manufacturers. This ranges from 'newsletters' and 'sales literature', which contains little or no technical information, to the more extensive 'technical literature', which contains specifications and detailed drawings. It also comprises advertisements in the professional journals, direct mail and technical information, a growing proportion of which is also supplied in electronic format via the internet and on computer disks provided free of charge to specifier's offices. This is 'one-way' communication that relies on the specifier becoming aware of the information and making contact with the manufacturer to obtain further information prior to selection. The second method of communication is through the trade representative (also known as a 'sales representative' or a 'technical representative'), a person who forms an interpersonal link between the manufacturer and the specifier's office. Such a person is referred to as a 'change agent' in diffusion of innovations literature (e.g. Rogers, 1995), a role that provides the opportunity for information exchange between manufacturer and specifier, and whose ultimate aim is to get the specification. Both forms of communication have to pass through some form of gate or 'gatekeeper' to get into the designer's office. The trade representative must get beyond the office receptionist in order to see the potential specifier, and direct mail must survive any filtering by the partner when the mail is opened if it is to stand any chance of finding its way to the individuals who actually do the specifying.

Research published in 1971, *Architects and Information*, by Jane Goodey and Kate Matthew, looked at how architectural firms handled technical information, with a view to designing technical information that might appeal to specifiers. At the outset of their research, it was known that a lot of effort was being wasted because information produced (both technical guides and manufacturers' trade literature) was not being read by architects in practice. They concluded that this was because it was not provided in an accessible form for specifiers and suggested a number of design guidelines to be followed by manufacturers that in the event were largely ignored, despite the existence of British Standard 4940 (BS, 1994). Their report drew on earlier work by Dargan Buliivant (1959) and was followed by other related work by Mackinder (specification) in 1980 and Mackinder and Marvin (use of information in decision making) in 1982. More recently, research into the specification of building products has been published by the Barbour Index (1993) and Emmitt (1997).

Specifier's awareness

Marketing strategies adopted by the different building products manufacturers differ. The most recent of the reports published by Barbour Index (1996) *Communicating*

with Construction Customers – A Guide for Building Product Manufacturers used three case studies to highlight different manufacturers' communication strategies (Barbour Index, 1996:22–26); in Rogers' terms, these can be classified under two communication channels, mass media and interpersonal channels.
Mass media channels comprise:

1 advertisements in journals;
2 direct mailing of trade literature; and
3 entries in product directories.

Interpersonal channels include:

1 trade representatives;
2 technical support;
3 continuing professional development (CPD) seminars;
4 exhibitions, seminars, and product displays.

However, the important channel here, because of the technical support that this person can provide, is the trade representative, who may either respond to telephone enquiries or visit the architect's office. This technical support provides additional knowledge about a product after initial awareness. While CPD events and exhibitions rely on potential specifiers actually attending the events, for medium and large offices, this is achieved by representatives of manufacturers putting on events at architects' offices, often at lunch-times when some refreshments may encourage an informal atmosphere conducive to a 'soft sell'. The result is that sole practitioners or small practices often miss out on this channel of communication. There are therefore two factors affecting the routes by which knowledge of innovations may be transmitted to designers. One is the marketing policy adopted by a manufacturer, and the other is the relative size of the practice.

The strategies reported in the Barbour reports are similar to those identified in three interviews carried out with different building-product manufacturers. The companies' marketing managers were interviewed and asked how they raised the awareness of potential specifiers to their products. All three had the same strategy, their marketing campaigns relying on advertisements in the trade journals and direct mailing, supported by their trade representatives who introduced the product (with samples and trade literature) when visiting architects' offices. Trade literature was also sent out when a specifier responded to an advertisement in the trade journals, which was also followed up by a visit from the trade representatives. These interviews were carried out with the marketing managers of large, successful companies who had the resources available to employ a wide range of marketing techniques. Many other, smaller, companies do not employ trade representatives who are an expensive resource, and so have to rely almost entirely on paper information to bring about awareness among potential customers.

Mass media channels

Journal advertisements

Building-product manufacturers may advertise their products in professional journals (*RIBA Journal*, *Journal of the AIA*), in commercial journals aimed at the professional audience (*Architects' Journal*) or in product journals such as *Building Products* and *What's New in Building* that are issued free of charge to potential specifiers. The advertising strategy usually includes paid advertising space and less expensive press releases, usually presented as a technical article in the journal. Both forms of advertising are used with the intention of raising specifiers' awareness of the company's product. If this happens, the specifier becomes aware during a 'passive phase', i.e. not actively looking for information about building products.

Once the specifier has become aware of a building product, he or she can either telephone the company to ask for their literature or can fill in a 'reader reply card' (if provided by the journal) to request further information by post. Either way triggers a mailing of literature to the architect's office that is often followed by a telephone call from the company's marketing department or a visit from a trade representative (if the manufacturer employs them).

The product directory

Published annually, the product directories are issued free of charge (or purchased for a nominal sum to architect's offices). Typical examples in Britain are the *Architects Standard Catalogue* (*ASC*), the *Barbour Index*, the *RIBA Product Selector* and *Specification*, while in America, *Sweet's* catalogue serves this vital role in architects' offices. These publications list building products by element, and to differing degrees include advertisements by manufacturing companies: *Specification* provides advice on how generic products should be used and specified. These are available in printed form and also electronically (e.g. *Barbour Index*), but companies have to subscribe to these compendia, and not all do so. They are intended as a source of reference for the specifier, so awareness through this medium could occur a long time after the first advertising campaign has finished. This form of awareness relies on the potential specifier looking through the compendium and then contacting the manufacturer for further details, i.e. the specifier has to be in an active phase. While the compendium will not draw attention to the newness of the product, in comparison to the advertisement, it may be perceived as new to the specifier who is looking through the directory.

Trade literature

The term 'trade literature' is used here to describe both the manufacturer's technical literature and sales literature, partly because it is sometimes difficult to distinguish between the two and partly because both are used to raise the specifier's awareness. Trade literature is supplied free of charge, but can enter the office in a variety of ways:

1. through unsolicited direct mailing, where trade literature is posted to all architects' offices listed on a database of addresses in the hope of raising awareness;
2. in response to a request from a potential specifier who has perhaps responded to an advertisement as noted above, i.e. after initial awareness of the product; and
3. through an update of information by a technical library maintenance service that is used by some offices.

In the first two cases, the material may come direct to the specifier or may have come into the office through the office's own librarian. In those cases where the material has come into the office library, whether through action by the office's own librarian or because of some library maintenance service, awareness relies on a specifier looking in the library for information (unless there is a policy of circulating it round the office first).

Trade literature has a different purpose from journal advertisements or entries in the product directory. Journal advertisements and entries in directories are only designed to raise awareness of the product; the specifier must send off for further information to enable it to be specified. Although trade literature can be a source of initial awareness, landing on the specifier's desk at the right moment, its primary function is to provide the specifier with enough information to be able to specify the product or certainly enough information to contact the company and invite the trade representative into the office. Here, we can distinguish between technical and sales literature because, while the former will facilitate specification, the latter will not. Differences in the design and intended purpose of literature will depend on individual company's marketing strategy, but there is nothing to prevent technical literature being used as sales literature.

Some companies produce no technical literature, their strategy being to raise awareness of their products though sales literature and then to respond to enquiries through a visit from their representative: such different marketing strategies imply different awareness channels. This is not necessarily a feature of the type of product being marketed. Companies manufacturing very similar products and therefore presumably aiming at a similar market may differ in this respect, one sending out full technical literature on request, the other not. The work reported above suggests that this is unlikely to be a successful strategy because of the reluctance of designers to 'waste their time' with trade representatives.

Information is also available in electronic format, on computer disks, on CD-ROM and, since 1998, even freely available on the web. All these forms offer benefits in speed of response from a specific manufacturer for additional information and save space taken up by trade literature in the office library. Such systems cost the architectural practice money and, according to the Barbour Index (1996:28), are only used by a very small proportion of specifiers (3 per cent). More recently, commercial organisations have been promoting on-line information via a common internet search.

With the steady growth in the number of architectural practices who use Computer-aided Design (CAD) software packages, some manufacturing companies have responded by providing technical information to specifiers on floppy disk, usually supplied free of charge. They hope that by providing typical details of their product in a form that can be easily imported into the architect's drawings, there will be a greater chance of adoption. This is essentially a new form of technical literature.

All of these systems rely, to a greater or lesser extent, upon the specifier actually noticing the information, through reading a journal, looking for information in a directory or by the trade literature landing on his or her desk at an opportune moment. Thus, the mass media channels rely on hitting the specifier at the right time, when the individual is engaged in the task of detail design. This assumes that no one remembers this kind of information, which may happen in some cases. A study of communication patterns in engineering firms found that there are those within organisations who act as sources of information because of their ability to keep up to date with developments. This is not necessarily information on new products but may be scientific findings that affect the kinds of design in which the firm is engaged. It was found that such people tend to be the centres of communication networks within the firms and thus may be important disseminators of information. In management literature, the function of taking information from outside the organisation and then passing on relevant aspects of it, that is in a filtered form, to those within the organisation is known as a boundary role (e.g. Allee, 1997). In design firms, a similar function may be performed by the technical partner: someone whose responsibility is the technical quality of the designs. In the gatekeeping research, it was found that these people were a source of technical information within firms but also acted to limit their employees' awareness of innovative products.

Interpersonal channels

The main interpersonal channel of communication is via the trade representative. The term trade representative is used to cover both sales representatives and technical representatives. Although not employed by all building product manufacturers, they provide an important link between the manufacturer and the unit of adoption, the specifier's office, and are classified by Rogers as the change agent;

> A change agent is an individual who influences clients' innovation-decisions in a direction deemed desirable by a change agency. A change agent usually seeks to secure the adoption of new ideas ... (Rogers, 1995:335)

The trade representative has a dual role, employed both to raise the awareness of the specifier to the company's products and to provide further information once the specifier has become aware of the new product through some other source, such as an advertisement in a journal.

Rogers (1995:337) has identified seven roles in the process by which an innovation is introduced to the potential adopter by a change agent. The sequence of change agent roles is to (1) develop a need for change, (2) establish an information-exchange relationship, (3) diagnose problems, (4) create an intent in the client to change, (5) translate an intent to action, (6) stabilise adoption and prevent discontinuance, and (7) achieve a terminal relationship. It is clear from the observation of specifiers' behaviour, described above, that roles 1 and 4 are not needed by the change agent in the design office. Specifiers have shown themselves resistant to innovations unless they already have a need created by a particular problem when they will be actively looking. However, trade representatives act as if they were unaware of this and continue to assume roles 1 and 4, but before they can carry these out, they have to get their 'foot through the door' of the architect's office, and this is an important stage in the process. The trade representatives, or change agents, will attempt to communicate with potential specifiers in the architect's office, either by visiting the office or by telephone calls. Although they form an important link or bridge between the two different social systems, the professional office of the architect and the commercial world of the manufacturer, the very difference in their social position is a problem. Rogers has discussed the problem of compatibility between change agents and their clients, which he has called the heterophily gap (Rogers, 1995:336), where the change agent is often perceived as having low credibility by the potential adopters. In other diffusion research, difficulties in communication are related to different values and different levels of education, in Rogers' examples, the change agents being more highly qualified than the clients. Unless the trade representative is also a qualified architect, it is the potential adopter who is likely to be more highly qualified than the trade representative and will certainly consider himself to be so. Thus, while the heterophily gap exists, the status levels are reversed compared with those described by Rogers. This has an effect that is different from that described by Rogers where potential adopters sometimes felt reverential towards the change agent who was assumed to 'know better'.

Apart from the Walton Markham Associates (1981) telephone survey where architects reported that trade representatives were 'a necessary evil', there was no other published investigation of this relationship. A small pilot study was carried out in which 10 trade representatives were interviewed during visits to an architect's office. The representatives were selected simply by the fact that one of the authors had a few minutes' spare time to see them over a 4-week period. Of the 10 representatives interviewed, only one was a qualified architect, a strategy adopted by this firm because they were aware of the problem noted above. Four said that they had a qualification in marketing, whilst the other five claimed some experience of the building industry. Those with a marketing background had previously sold a variety of products, from car components to chocolate bars. Although the interviews were not designed to be representative of all trade representatives, they did indicate the heterophily anticipated. The representatives were candid, but did report a reluctance on behalf of specifiers to look at their new product range unless they happened to catch them when they were detailing and specifying buildings where such products may be of use to them. These findings were also observed in a separate research project (Emmitt, 2001b) that asked partners and trade representatives for their views of one another and is summarised below.

The partners' view

Partners were asked about their relationship with the trade representatives. In the five very small offices, the partners declared themselves to be much more approachable, compared with those in the other offices. However, because these practices worked on smaller projects, the trade representatives paid them less attention. This was illustrated in one instance where a very small practice and a medium to large practice were located 100 m from each other in the same street. The very small practice was rarely visited by trade representatives, despite the fact that they had to walk past the door to get to the other practice. The larger practice complained about being pestered by the trade representatives, and the receptionists had been issued with instructions to politely deter them from visiting again: none who called without an appointment were entertained, and appointments were always to discuss products that related to a current project.

The experience of these two offices reflected a general pattern. The partners in the small offices said that they rarely saw trade representatives, whilst four of the five large offices had a policy of not seeing trade representatives unless they had been specifically invited in to the practice by a member of staff. Generally, trade representatives who called without an appointment were politely told by the office receptionist that no one was available to see them or were 'palmed off' with the office junior or student architect. The only exception to this was one partner in a medium to large practice who made an attempt to see trade representatives when possible, partly because he was interested in new products and partly because he said it provided a break from an otherwise hectic job.

Another factor in this was the architects' views of the standard of trade representatives. All of the partners complained about, and questioned the competence of, the majority. When asked to explain their reason for this view, they said that many of the representatives had very little technical knowledge, often described as knowing less about the products they were trying to sell than the architects who they were trying to sell to. However, these opinions were largely formed second-hand from conversations with the specifiers in their office rather than from personal knowledge of the representatives, who they rarely met.

It had been assumed that the office receptionist would act as a gatekeeper to incoming telephone calls from manufacturers and to trade representatives when they attempted to visit the architect's office. While there was no evidence to the contrary, the only supporting evidence came from two of the receptionists who said that if trade representatives called without an appointment, the best they could hope for was a few minutes of the office junior's time and noted that the representatives were 'rather a nuisance', seen as distracting them from more urgent matters. In addition to these comments, a receptionist in a small practice claimed that she dealt with the majority of telephone calls and trade representatives without reference to the partner or architects within the office because she knew when the staff did not want to be disturbed: she saw this as the duty of a professional receptionist and thought that all receptionists acted in this way. The partners were thus unaware of the extent of sales pressure.

In three cases where the reception areas were located on upper floors of buildings, a physical mechanism was employed in the form of an electronic entrance control system. Installed primarily for security reasons, it was also used by the receptionists to prevent people calling into the office without an appointment (trade representatives and contractors). All three of the receptionists said that they rarely let anyone into the office without an appointment (it was also very difficult to contact these three offices by telephone). In these three examples, the entrance control mechanism prevented trade representatives from calling at the office speculatively because they simply could not get into the building.

A view from the trade representatives

A further five interviews were carried out with experienced trade representatives; each of the representatives had been selling building products to specifiers for over 10 years and therefore could offer an experienced view of the process from their perspective. The representatives were selected because they were good at getting through the gate, in contrast to many more who were not, and this will have influenced the views received. These particular representatives were keen to use the meetings with architects for feedback on existing products and for views on new product development because they viewed themselves as an important link in terms of research and development of new building products. This may be a reflection of their companies' policies. All five complained that architects were very difficult to 'sell to', compared with contractors, but took it as the challenging part of their job. They were aware of the personal collections of trade literature used by architects, and all claimed various techniques for getting their information into these. The representatives were aware of the 'foot through the door' problem that existed when visiting architectural practices but noted that it was much less of a problem with contractors, quantity surveyors or planners.

To overcome the architects' resistance, they marketed their products to other members of the building industry in an attempt to raise the architect's awareness (and get their products specified) by indirect means. Their trade literature and knowledge of their products were often communicated to the architects by pressure from, for example, the planners. They were aware of the influence of town-planning officers over the choice of external materials, the influence of the quantity surveyor where cost was paramount and the role of the contractor in changing products whilst the building was being constructed, known as 'switch selling' but referred to here as 'specification substitution'.

Three of the representatives explained the greater ease of communication with other professionals as due to arrogance on behalf of the architects, whilst the other two had concluded that architects were not interested in new products and so were immune to their sales techniques. The trade representatives felt that architects tried to put over an impression of being technically competent, when in fact they were often poorly informed about specific technical issues. This contrasts with the view of the architects who were equally dismissive of trade representatives. In fairness to these trade representatives, they felt frustrated because they believed that a closer working relationship between the manufacturer and the architect would benefit both (this was not a view shared by the architects who saw it as 'a waste of their valuable time'). Here, of course, we are comparing the views of a small number of competent representatives with designers' views of all trade representatives. Thus, this difference of view is not unexpected.

Four of the five representatives who marketed external materials said that it was just as important to get their information into the planner's palette of favourite products as that of the designer. For example, the representative who marketed artificial stone products said that he had been particularly successful in selling a large quantity of his company's product directly through planners working in one particular Local Authority planning department, because they recommended his company's product by name to architects submitting schemes. Mackinder (1980:143) also found architects who reported that planners had a tendency to recommend products by brand name. This is dubious practice from the point of view of both professional ethics and professional responsibility. The representatives also felt that they were successfully selling products to architects through the main contractors and, in some instances, the quantity surveyors, all of whom were easier to see and were perceived as more responsive to their products than the architectural offices.

Rogers (1995:350) found that the greater the homophily between client and change agent, the greater the success in securing adoption. Some manufacturers employ architects as trade representatives in an attempt to reduce the heterophily gap. For example, Ibstock was the first brick-making company to employ a trade representative (Cassell, 1990), a marketing strategy that has since been copied by their competitors. This approach is reflected in their sales figures, with 60 per cent of all orders from architects' specifications, which their marketing manager attributed in part to their trade representatives ability to 'get close to the specifier' or homophily with the specifier. Rogers also noted that change-agent success will depend upon the characteristics of the product innovation that the representative is attempting to get adopted. He commented that it depended upon how well the innovation fitted into the recipients' existing belief system, and there would be a parallel here in architecture. The representative trying to promote high-quality facing bricks may stand a better chance of seeing the architect than if he was trying to promote, say nails, since the former is central to architects' interests, while the latter is not.

Rogers concluded that one of the most fundamental factors in the success of the change agent is the extent of change agent and client contact; the greater the face-to-face contact, the greater the likelihood of adoption and subsequent diffusion. Thus, the most important stage is to initiate contact with the potential specifier, i.e. the trade representative has to get his 'foot through the door'. Some trade representatives may already have established a social relationship between one or a number of specifiers in an architect's office that may be long established, rather than created for a single occasion. It is likely that these representatives will find it easier to introduce new products than those representatives who are trying to establish a new relationship with the office.

Technological gatekeeping

There are two questions relating to awareness raised in this section so far. First, what happens to the trade literature when it enters the architect's office, does it get passed to the specifiers, or is it filed in the office library? Second, how successful is the trade representative at establishing a relationship with the architect's office? The answer to both depends upon management strategies adopted by offices as much as by the marketing policies of the manufacturers because offices are not simply passive recipients of the attentions of the manufacturers. Offices operate some form of filtering process, gatekeeping mechanisms that operate on these two communication channels to limit the flow of information. Trade representatives may be regarded as necessary evils, but the evil attribute is as significant as the necessary and affects their ability to get their message to architects. Also, the trade literature that comes into the office may be regarded as much as a nuisance as a welcome source of information. Because there are two kinds of message carriers, there are two different kinds of gate operating that need to be considered separately. They may be formal gates, consciously managed by the practice or simply informal barriers.

Despite the existence of BS 4940:1994 *Technical information on construction products and services*, the fact that trade literature is supplied free of charge and that the specifier relies heavily on it, it does vary in quality: as such, it needs to be controlled by the specifier's office. Literature needs to be controlled, both to avoid information overload, a state achieved when an individual or organisation receives more information than it can handle leading to some form of breakdown (Rogers, 1986), and to ensure that it is a reliable source of information. The way in which it is controlled is an aspect of office management policy usually applied at a senior level. This has the effect that the more junior members of staff in medium and large practices are shielded from excessive information or denied knowledge of potentially useful innovations. Whichever way one looks at it, this filtering practice must affect the uptake of innovations.

Information overload

A problem facing individual specifiers and managers of professional service firms is the vast quantity of information available to the organisation's members. This is a particular problem for the 'information rich' (Rogers 1986), professionals such as engineers and architects who need to balance the need to stay abreast of current developments with a manageable information base. With the advent of cheap publishing and the growth of digital information, the need to manage information has become a challenge. The volume of information has increased to such an extent that some form of specialised management structure is required to store, process and retrieve relevant information if a state of information overload is to be avoided.

It is tempting to view information overload as a new phenomenon, but it has been a cause for concern amongst professionals for some time. For example, in *The*

Principles of Architectural Design published in 1907, Marks suggested that architects should look at the trade catalogues left by travellers (the forerunner of the trade representative) on a weekly basis and then dispose of them after reading – an early example on managing the volume (if not quality) of technical information. The volume of information targeted at design offices has grown significantly since Marks' advice. Goodey and Matthew reported that offices were being 'submerged by a flood of literature' in their report published in 1971, a situation that has worsened. Studies in Japan (e.g. Bowes, 1981) and the United States of America (e.g. Pool, 1983) into information use by societies found an extraordinary growth in information yet a modest growth in the consumption of information, suggesting that individuals operate selective exposure.

The technological gatekeepers

Filtering of information is known as 'gatekeeping' in communication studies, a metaphor first coined by Kurt Lewin (1947), studied empirically by David Manning White (1950) and since developed into a field of research most commonly concerned with mass media studies. The gatekeeping metaphor can be applied to any decision point where information, provided via mass media or interpersonal channels, is assessed for onward transmission by a 'gatekeeper' (Shoemaker, 1991). The construct is present in diffusion of innovation literature, both explicitly (e.g. Greenberg, 1964) and implicitly (e.g. Rogers, 1962) and is present in literature concerned with knowledge acquisition, where the term 'technological gatekeeper' is used to describe an individual who attempts to control information entering the organisation from external sources (e.g. Leonard-Barton, 1995). Such individuals operate at the boundary of the firm, where they 'browse' information for its relevance to both themselves and their organisation's members, withholding, altering and transmitting information as it passes into the social system over which they have a certain amount of control. Thus, gatekeepers may influence the innovativeness of the organisation, simply by the type of information they allow through the gate.

For our purposes, the gatekeeper is someone who may withhold information as it goes from the sender, the building product manufacturer, to the receiver, the potential specifier. The partner and the office receptionist are potential gatekeepers, being the first gate encountered by trade representatives and trade literature. Both have to pass through a *gatekeeper* to get into the office. The trade representative must pass the office receptionist in order to see a potential specifier, and direct mail must survive any filtering by the partner when the mail is opened if it is to stand any chance of finding its way to the specifier. Mackinder (1980) found that partners in architectural practices were seen to have an influence over specification decisions, deciding on standard specifications and making decisions where new products were being considered. It is not unreasonable to assume that the partner of an architect's office may also seek to control trade information as it entered the office. In doing this, the partner is likely to ask questions such as: 'What do the members of the office need to know?' and 'What do they already know?', i.e. the gatekeeper will operate a set of routine questions when deciding what information can pass through the gate, a phenomenon described by Windahl et al. (1992).

Architects have also confirmed that trade literature entering their office was filtered by senior staff, often the senior partner, then filtered further as it was passed to less senior architects and technicians within the office. Therefore, information has to pass through a series of gates, controlled by different gatekeepers, before it reaches the specifier who may need it.

Gatekeeping research

A small research project, based on qualitative interviews with 15 partners of architectural practices and their receptionists, plus interviews with five trade representatives (to add a degree of balance to the research), was carried out to investigate how and why trade literature is managed as it enters the architect's office. The objective was to try to identify the gatekeeping mechanisms as follows:

1. What is the office policy towards trade literature, and how is it filtered, disseminated and stored in each office? This enabled the second and third more general questions to be addressed.
2. What type of information is let through the gate, and why?
3. To what extent do these policies affect the architect's exposure to new building products?
4. By comparing the different offices, it was also hoped to consider how the mechanisms employed were influenced by office size.

No attempt was made to measure the reduction in the amount of trade literature affected by this process.

Interviews were held with the partners of five very small practices (one to five designers), five small practices (six to 10 designers) and five medium to large offices (11 designers or over). All of the offices were private architectural practices, selected from two different geographical areas in the RIBA's list of practices, a metropolitan area and a large county town. Offices in each of the three size categories were sent a letter asking for their co-operation, but the response to this was poor. When appointments were then solicited by telephone, this proved an equally difficult task since in over half of the calls to the selected offices, it was impossible to get past the receptionist to talk to the senior partner. Eventually, five acceptances in each category were obtained. Thus, the architectural practices were, in part, self-selected and could hardly be described as a random selection. Interviews were only arranged in offices where it had been possible to penetrate the gatekeeping mechanisms that we were trying to examine. This is an important point to make because it may have affected the results. We may assume that the interviews were conducted with offices that were operating less stringent gatekeeping mechanisms.

The private architectural practices that were interviewed were engaged on a variety of work, although each tended to specialise to a degree. Of the very small practices, four of the five were engaged equally on refurbishment and new build, whilst the fifth concentrated almost entirely on new build. The small practices concentrated mainly on new build work, one claimed to specialise in housing, whilst the other four claimed expertise in retail and commercial work. This pattern was similar in the medium to large offices who concentrated on larger commercial and retail developments, but who also took on small house extensions.

None of the practices interviewed used electronic (CD) trade literature systems, although partners in two of the medium to large practices said that they would if the cost was reduced. The RIBA librarian service was employed by one of the very small practices, three of the small practices and two of the medium to large practices. This meant that RIBA-approved literature was introduced to the library, bypassing any gatekeeping mechanism regardless of its perceived usefulness to the office. Those that did not use the service had allocated one of their staff to keep the library up to date, by filing the (previously filtered) literature.

The partners turned out to be ruthless in their handling of trade literature, as the majority was thrown away during the morning ritual of opening the mail. When asked to estimate the approximate amount of trade literature that was retained and passed to other architects in the office, estimates ranged from 1 per cent to a maximum of 20 per cent. Thus, 80 per cent or more of all incoming trade literature was thrown away by the partner. When asked to explain how they decided what to keep and what to throw away, the selection criteria employed appeared not to be particularly objective because they had limited time in which to assess the information. On average, the partners said that approximately 5 minutes was all the time that they spent on this task, and most said that the literature had only 3–5 seconds to command their attention, otherwise it was thrown away without even looking past the front cover. Two factors emerged that appeared to lead to retention: first, was the literature relevant to the type of work of the office, and second was whether it appeared to contain enough technical information to make it worth keeping. If the literature was perceived to be of little use to the office and/or the literature did not contain 'good technical information', it was thrown away. Thus, the glossy brochures containing photographs and little technical text were discarded.

However, regardless of its perceived quality, literature produced by companies that were familiar to the office was more likely to be let through the gate than that from companies less well known to the partner. This was seen as a risk-management technique and was consistent across the full sample, regardless of office size. There were no strong views about the physical size of literature. Interviews revealed an underlying level of dissatisfaction with the general standard of trade literature.

Further gates

In the very small offices, the partners also acted as specifiers because they claimed to retain a 'hands on' approach, dealing with design issues in addition to running the business side of the practice. Here was a difference between the very small offices and the others because the partners in larger offices were primarily concerned with the business side of the practice; none of them was personally involved in product selection, despite the fact that they claimed to keep a close involvement with all of the jobs. Of course, we cannot be sure of this, because Cuff (1991:20) noted the differences in partners' actual behaviour and their own self-image. However, their reported behaviour was reflected in office procedure, where trade literature was passed from the senior partner to a partner or associate who was responsible for technical issues. Thus, once past the first gatekeeper, the information faced further filtering in the small and medium to large practices by a less senior member of the office: the estimated reduction in literature at this stage was a further 50 per cent, based on an objective (and more time-consuming) assessment of the literature.

One of the medium to large practices employed a very elaborate gatekeeping system. Any trade information let through the gate by the partner went straight to a senior architect (associate level) whose job was to deal with technical matters in the office. If the product was perceived to have a degree of benefit to the practice, he then contacted the manufacturers directly for further information. Once this information was received (usually delivered by a trade representative), the literature was either discarded, if unsuitable, or presented to a partner's meeting and, if deemed suitable, recommended for adoption onto the practice's list of approved products. The reason for adopting such a complex system was founded on concern about product failure and the threat of legal action that may ensue, i.e. it was a risk-management system. All staff had to select from the office palette of approved products and were indirectly being restricted in their awareness of new products by the gatekeeping process. There is no indication of how common this particularly elaborate process might be. It would only be feasible in a reasonably large office. Nevertheless, the general principle of restricting the kind of products that employees may use in order to limit risk is very common, although not all will apply the restriction at this 'awareness' stage.

Office library

Once the trade literature had passed through the various gates, and a decision had been taken to keep it for future reference, it was placed in the office library. In three of the medium to large practices and one of the small practices, the information was circulated to the staff before being placed in the library. In another of the medium to large practices, a list was circulated once per month, giving both the name of the manufacturer and the product name of information that had recently been placed in the library (the literature itself was not circulated to the specifiers). This represents a modest attempt to improve the staff's awareness of new products.

Approximately half of the sample of offices said that they tended to keep information in their office library that 'might come in useful', whilst the other half only kept information that related to current jobs in an attempt to reduce the amount of space taken up by the literature. One of the very small practices did not keep any trade information; it was all thrown away, relying on the *RIBA Product Selector* for product information when needed; this was then placed in the file specific to individual jobs for future reference. Another of the very small practices had recently thrown out the majority of trade literature and only kept a few 'essential' items to reduce the amount of space taken up.

The specifier's role

Although we have considered the formal gatekeeping mechanisms operated by the office, specifiers themselves operate their own gates, and it is also important to look at the process from the specifier's perspective. In 1981, the *Architects' Journal* (August 1981:380) summarised the findings of a telephone survey, conducted by Walton Markham Associates Limited, a market research company based in London that asked 100 architects how they obtained information about building products. Their comments were that:

1 they obtain product information from journals and by sending off reader enquiry cards for selected products, i.e. initial awareness through the paper literature (advertisements and technical articles);
2 that direct mail 'bombardment' was seen as 'a nuisance to be borne';
3 they regarded promotional events and trade fairs as a waste of time;
4 they complained that technical literature lacked detail;
5 unsolicited visits to the office by trade representatives were not welcome; the architects said that they were not interested in being sold to, instead they would go in search of a product when needed;

6 all had a central (office) library of building product information, whilst 60 per cent also kept a personal file of information (palette of products).

The sample size was small, but the finding that 60 per cent kept a palette of favourite products supports Mackinder's (1980) earlier work. Perhaps the most important point was that architects search for information about a product when the need arises. Therefore, awareness is not simply passive: it is actively sought at certain times.

If a specifier does not identify a need to change a particular product from that previously adopted (from the palette of favourite products), his personal gate will be closed to trade information, and thus, awareness of building product innovations is unlikely. However, if the specifier is dissatisfied with a product used previously, he will identify a need to change and will actively seek an alternative, and thus, the gate will be open to information about building-product innovations. When a specifier wants a specific piece of information, he has two basic alternatives; he can try to get it from his fellow specifiers in the office (their previous experience and the records generated by the past work of the office), or he can search for it, i.e. consult manufacturers information. Work by Mackinder and Marvin (1982) found that architects may seek information 'only as a last resort'.

Conclusions drawn from the gatekeeping research

The effects of the sampling method that limit the general validity of the research exercise have already been discussed but it has provided some additional information that supports earlier work (Mackinder, 1980; Walton Markham Associates, 1981). It shows that architectural practices are very resistant to information about new building products, concerned about failure and the legal implications of their specification decisions.

To a large extent, the gatekeepers assumed that their experience and professional judgement would help to prevent the specification of unsuitable products by other members of their office, by limiting their choice of products from which to specify. While, in the theoretical model, gatekeeping is carried out to manage information overload, the interviews demonstrated that this was very much a secondary concern: risk management was the primary objective. Since approximately 80 per cent of the literature did not get through the first gate, the specifiers who actually selected building products were selecting from filtered information and so were affected by decisions taken by the gatekeepers because their access to, and awareness of, information was restricted.

Active and passive awareness periods

Obviously, the specifier must be aware of the building product innovation in order to consider it for adoption. Rogers has accepted that this is a chicken-and-egg problem – what comes first: the need to change or awareness of the innovation? In the diffusion of a medical drug (Coleman, 1966) the doctors did not seek information about the drug but gained knowledge of it from advertising and sales-persons, thus acting in a passive manner. Only when they were aware of it did they seek additional information about it. But we have seen that awareness from trade literature is restricted and that, according to the Walton Markham telephone interview, and the case studies reported earlier, specifiers search for information only when the need arises.

Communication is a two-way process, with sharing of information between the sender and the receiver. On the one hand, there is a constant and rapidly increasing volume of information, in both paper and electronic format, which is targeted at architects' offices. On the other hand, the specifier will require specific information for specific projects at different times. Bullivant (1959) noted that the architect's need for information is directly related to the stage, or timetable, of the particular project, and therefore, he or she may only be responsive to information about products if exposed to them when actively involved in producing working drawings. Thus, there may be some periods when architects will be receptive to incoming information and during which they will actively seek information and others when they are engaged on tasks unrelated to product selection, a passive period during which the information is likely to be ignored, i.e. specifiers will operate their own gates (see Figure 8.1).

Rogers (1995:164) has argued that individuals will seldom expose themselves to messages about an innovation unless they feel a need for it. Furthermore, even if these individuals are exposed to such innovation messages, there will be little effect unless they perceive the innovation as relevant to their current needs and consistent with their existing attitudes and beliefs. This is known as selective exposure (and selective perception) and implies that the need for an innovation will usually precede awareness of it. According to Rogers, awareness is related to individual level of attentiveness at the time of exposure to the innovation. Presumably, the specifier's level of attentiveness is higher when actively engaged in detail design, than in some other stage of the design process, as previously discussed.

There is an important issue for manufacturers here. The facts as presented are that manufacturers constantly bombard specifiers with information in the hope of hitting them during their active phase – at least a phase when they will be active for this particular kind of product. The gatekeeping experiment demonstrated that this is an extremely inefficient way of operating because, even if the specifier was prepared to consider this kind of product, the information may not reach him. Building product manufacturers who simply bombard the architect's office with direct mail obviously

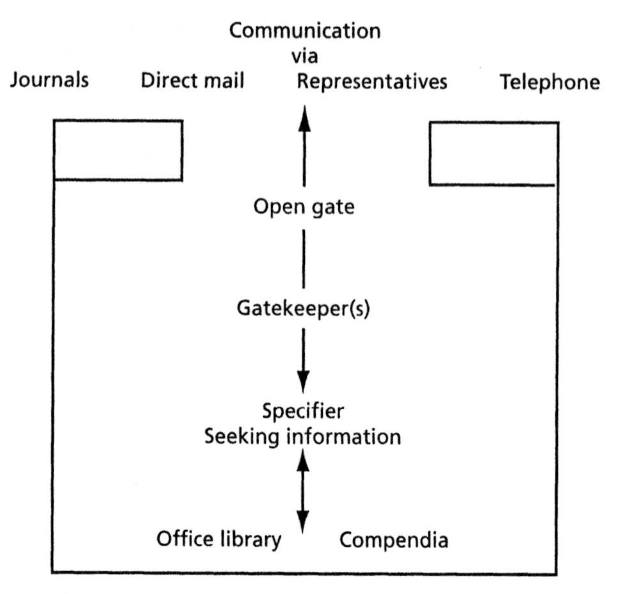

The 'active' specifier

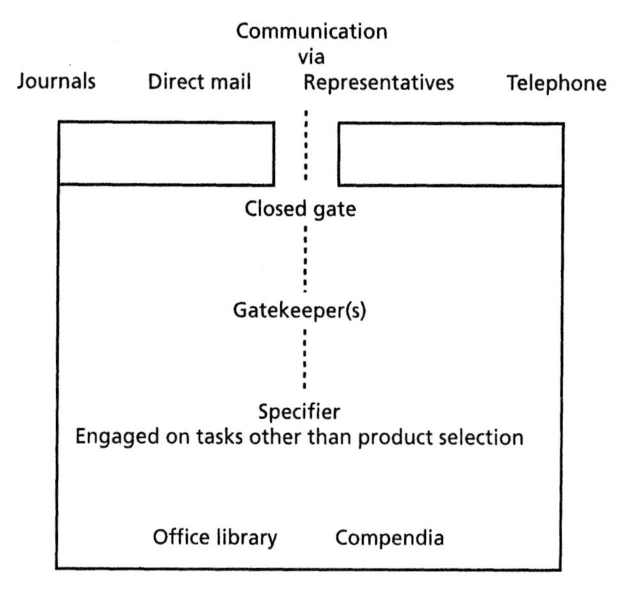

The 'passive' specifier

Fig. 8.1 Active and passive awareness period

hope that the specifier will become aware of the innovation by accident (due to random or non-purposive activities), or that they will be lucky, and the information will reach the specifier when he or she is actively searching for an alternative product.

Should this fail, they hope that the direct mail literature will be filed in the office library (or the specifier's personal library) for possible reference at a future date. Although trade literature may have been in the office library for some time, it is still a potential source of awareness.

A more efficient method of working might be to reach the architect during the passive period. This is not going to be achieved through mailing, and the trade representative is not going to get through the door in the normal way, which leaves only the Continuing Professional Development (CPD) mechanism. CPD events are often sponsored by manufacturers who hope to promote their own products, although the ostensible reason for these events is to deal with some technical issue associated with the product type. In such an event, potential specifiers are brought to a state of receptiveness by concentrating their minds on an issue that may not be directly related to the particular product. A group of products and associated issues might be discussed and while the associated issues might be concerned directly with design, they might equally be concerned with something quite different, such as construction safety.

As an example of the latter, lead paints have not been used for a long time now because they are poisonous. However, many other products, especially those with volatile solvents, may be toxic in application, and concern for health and safety issues has focused the attention of manufacturers on developing new products that are not toxic. The way in which these formulations have been developed and the implications of this will not be apparent to the specifier, nor perhaps of special interest to him, and yet, it is in manufacturers' interests to make the specifier aware of the improvements as a means of encouraging the specification of their products. This is an aspect of trade information that can perhaps best be conveyed during some CPD event when, away from the pressures of immediate design and production problems, and perhaps away from the office itself, the specifier will be able to give his mind to such an issue.

Resistance to 'new' products

Although building-product manufacturers use both mass media and interpersonal channels to communicate knowledge of their products to the specifiers, offices operate gatekeeping mechanisms that restrict the amount of information reaching specifiers. Those in the office who actually selected building products were not involved in the gatekeeping process but were clearly affected by decisions taken by the gatekeepers. Thus, their exposure to, and awareness of, building product innovations is restricted. Because of the gatekeeping mechanisms in place, it would appear that the trade representative may only enter the office by invitation, i.e. once the specifier has already identified a need, and not before. As such, the trade representative may have limited impact as an agent of change. He or she may also have a limited role in raising awareness of 'new products'. However, once inside the specifier's office, there may be an opportunity to discuss other products produced by the manufacturer in the hope of specification at a future date.

Specifiers may operate their own gatekeeping mechanism, closing the gate to information when engaged in tasks other than product selection, a passive period, and opening the gate if his palette of favourite products is unable to solve a particular problem, an active period during which information about building products is sought. This is clearly demonstrated in the case studies reported in Chapters 9 and 10. The manner in which the specifier becomes aware of information about a building product innovation is important because it is the first step in the innovation-decision process.

Where manufacturers, and often their trade representatives, were known to the practice, the partners were more relaxed in their attitude to information about new products. By way of contrast, the sample were very cautious when confronted by information from a manufacturer who was new to them: three of the partners felt that they 'had a duty' to investigate new products but did not have the time to pursue it, whilst the others said that they did investigate new products, but only when the need arose. This observation supported the earlier work by Mackinder (1980), where one-third of her sample noted that it was office policy to avoid the use of anything new if possible, preferring to stick to building products that they were familiar with. Her sample did, however, recognise that new materials and components needed to be monitored in case there were any advantages in terms of cost or performance.

Three of the sample, all from very small offices, said that they would not use products that had been used by famous architects, because they were perceived to carry a greater risk of failure. In contrast, one partner from a small office said that it would influence his decision to investigate it further, whilst there was no strong view from the remainder. All of the sample said that they had to protect the excess limit on their professional indemnity insurance: the more claims the higher the excess, and so 'safe (familiar) products' were preferred. This was summed up by one partner, who said 'We introduce new products to suit our, or our clients', circumstances not because a manufacturer wants us to. What's the rush?', a comment that supported the postal questionnaire respondents' desire to wait. It was this fear of unfamiliar products and manufacturers, perceived as risky, that had led to the employment of the gatekeeping mechanisms in the offices so that the literature did not get to the specifiers in the design office. This reported resistance to 'new' products was also recorded in the observations reported in Chapters 9 and 10.

Monitoring the specification of buildings

We start this chapter with a brief overview of research methods that could be employed to monitor the uptake of building products in the specifier's office. This is followed by a description of a research programme that monitored the uptake of building products in design office over a 3-year period. This helps to highlight what happened to information as it entered the design office, how specifiers became aware of the products, why they selected or rejected them, and the pressures to change their initial choice. This case study helps to highlight the complexity of specifying buildings while also identifying a number of issues that must be addressed in the pursuit of design excellence.

Awareness of 'new' products

Rogers has criticised diffusion research for the manner in which data have been collected in the majority of diffusion studies. The research has been based on retrospective data collection, i.e. an innovation that has been adopted and diffused is then traced back to examine the reason for its adoption, usually by interviewing the adopters some time after their decision to adopt the innovation was made: this reliance on the memory recall of the adopters was referred to as 'after-the-fact data gathering' by Rogers (1995:106). Obviously, this method of data collection also makes it difficult to research innovations that failed to diffuse because there is less information available (fewer adopters to ask), hence leading to a focus on successful innovations. Ways in which this can be overcome have been proposed by Rogers (1995:106–109), from data gathering whilst the diffusion process is still underway to the study of unsuccessful innovations. He has also suggested: that researchers need to acknowledge that rejection, discontinuance, and reinvention occur frequently during the diffusion process; that the broader context in which the innovation may diffuse should be investigated; and that the motivation for adoption needs to be addressed.

In addition to the information gathered by the postal questionnaire, evidence of what actually happens during the specification process was required because it is the innovation-decision-making process that is important. A detailed analysis of drawings and schedules would provide a list of products actually specified, while variation orders issued during the contract would confirm any changes to specified products that have occurred during the contract. However, it would not provide any evidence of how and when the specifier became aware of the building product, nor would it provide any insight into a specifier's innovation-decision process, because there is no reason to record this in the normal course of work. Furthermore, it would be both time-consuming and extremely difficult to identify which products were new to the specifier without resorting to an interview.

Previous research into specifiers' behaviour had relied on asking specifiers what they did, and/or asking them to record their decisions. Such approaches are valuable, but prone to difficulties. First, professionals tend to portray themselves as they wish to be seen (Ellis and Cuff, 1989), and their account of how they act may not necessarily reflect what they do. Second, is the ability to remember and recount all the steps in the process, something other researchers have found designers to be rather poor at (Yeomans, 1982). Third, asking specifiers to record their own behaviour in diaries naturally raises their awareness to specification decisions and may influence their behaviour.

As discussed above, it would appear that the adoption of products that are 'new' to the architect's office is likely to be a rare event in comparison to the large number of building products selected for a particular building project that are already known to the office. A method was required, therefore, that could trace a number of building product innovations from initial awareness to adoption (or rejection) which did not rely on memory recall, but which was based on monitoring the process. Participant observation was the method used.

The office

The observations were conducted in the design office of a well-established architectural practice that concentrated on industrial, commercial and retail developments. Approximately 90% of their work was new build and 10% refurbishment projects. The practice had gained a number of design awards but had never been featured in the architectural press. The practice has a good reputation for service and delivery and had recently introduced a number of management techniques aimed at improving the delivery and quality of their buildings. The staff structure remained constant throughout the data collection period; three members of staff left and were replaced with similarly qualified persons, discussed below.

The part of the drawing office in which the experiments were conducted was open plan with seven specifiers plus the author, comprising three qualified architects, three architectural technicians and an architectural student. The author (S.E.) was employed in a managerial role, and therefore, the tendency to ask 'why' certain decisions had been taken was a part of the daily routine and was unlikely to influence the behaviour of those being observed. The design office provided the opportunity to monitor both the adoption and non-adoption of building product innovations as decisions were taken by specifiers.

One of the difficulties of observing the design process is the time that this takes, although participant observation has been employed by researchers like Cuff (1991). This was also possible here, and a diary of adoption was kept over a 40-month period. During this time, an attempt was made to record any event that affected the process of adoption. One of the problems, however, was to ensure that this was a practical proposition. To ensure this, nine products were selected for monitoring and the first stage of the process, that is, the initial knowledge stage was deliberately started by the observer. The experiment was begun by requesting trade information about a number of products by using the 'Reader Reply' card enclosed with one of the free 'product' journals received by the office. Products were selected to cover a variety of uses:

- Product 1. A uPVC product to be used with coping bricks for brick walls.
- Product 2. An external decorative facing panel.
- Product 3. A uPVC cavity closing system incorporating a thermal barrier.
- Product 4. Reinforcing mesh for retaining soil on sloping sites.
- Product 5. Timber flooring system for use as an internal finish.
- Product 6. A product to assist the pouring of concrete on site.
- Product 7. Thermal insulating building blocks.
- Product 8. An internal partition walling system.
- Product 9. An electronically controlled door lock.

There was standard office procedure, in which trade literature received by the office was passed from the senior partner to the technical partner on a daily basis and circulated to the staff at approximately 4-week intervals, before it was filed in the office library. The technical partner's job was to assess the literature and only circulate that which he thought would be both useful and technically acceptable, i.e. he acted as a technical gatekeeper, controlling the flow of trade literature to the specifiers in the office. To enable the observer to monitor the effect of this, the receptionist (who opened the mail before passing it to the senior partner each morning) was asked to record the date when information about the nine products was received in the office.

Trade literature relating to all nine building-product innovations was received over a 2-week period, but the only literature to reach the technical partner related to

Products 1, 2 and 3. Two-thirds of the information requested had failed to get past the senior partner during the morning ritual of opening the mail; it had been thrown away. This posed a slight problem for running the observation.

The significance of this filtering proces is discussed below, but it hindered the research by reducing the sample to only three products. Therefore, the information was requested again by telephoning the manufacturers directly. Since it was likely that the senior partner would again throw the information away, the receptionist was asked to intercept it and pass it directly to the technical partner, thus bypassing the senior partner. Information on five of the six building products was received and placed in the technical partner's 'in tray' by the receptionist. Information relating to Product 9 was never received. Thus, after 5 weeks, trade literature relating to eight of the nine building products had been received.

Two gates

Earlier, we discussed the filtering of trade literature, and this observation helps to illustrate its complexity, confirming evidence of more than one gate through which technical literature has to pass before it is retained within the office library of information.

The first gate

The senior partner, when asked about his gatekeeping actions, could not recall the particular examples, but did confirm that 'the majority' of trade literature received through the post every day was thrown away. He said that his decision to keep or reject information was 'something done quickly and based on experience'. Literature that looked as if it was well produced and might be of use to the office tended to be passed to the technical partner; the remainder (approximately 2/3) was thrown away. This reduced the physical amount of literature passed on to the technical partner and the amount of time he spent assessing it. The technical partner's job was to limit the office's exposure to risk by controlling products available to the specifiers in the office through this means. Thus, a product perceived as 'risky' should be rejected before reaching either specifiers or the office library. Although the senior partner delegated the task of vetting trade information to the technical partner (as described in the office manual), he acted as a gatekeeper to reduce information overload (which was not mentioned in the manual) before it reached the second gatekeeper.

The second gate

At the start of the 6th week, the technical partner circulated a tray containing his approved trade literature to the specifiers in the office. This contained information from 27 different manufacturers, ranging from a four-page brochure on handrails to three binders containing a variety of products. Six of the brochures related to the pre-selected sample, so the technical partner had rejected two of them based on his assessment of the trade literature. These were Products 3 and 6.

When asked about this process, he said that Product 6 was of interest to him because he thought it would save time in construction, and he had requested additional technical information by telephone. Curiously, the information was placed in his personal collection of literature and not circulated to the staff, nor filed in the office library. Despite his personal interest in this product, it was never specified by the practice and was still in his personal collection of literature at the end of the experiment. When questioned about this, the technical partner said that he fully intended to use the product, but had not had the time or the opportunity to investigate it. Having said that, he still thought it was 'a good product with potential and would use it at some point'. The important point here is that he had, unwittingly, prevented the specifiers in the office from seeing the information, and therefore, it could not be specified by them unless they became aware of it via some other source.

The second rejected product, Product 3, provided an example of a product that, as far as the technical partner was concerned, was launched too early. It had been launched to anticipate the introduction of the revised Building Regulations to reduce the effect of thermal bridging in cavity wall construction at window and door openings. Although the technical partner thought that it was an 'interesting product', he was concerned about its use because it was made of uPVC. Because he thought it 'too risky to use', the information was thrown away. Despite this, the product reappeared during the monitoring period (see below).

Six products were perceived as being potentially useful and, more importantly, 'safe'. Although Product 1 was also made from uPVC, it was not viewed with the same suspicion as Product 3 because it was for use on boundary walls rather than cavity walls of a building. This was regarded as less of a risk to the architect's office by the technical partner.

Information about the six building product innovations that had passed through the second gate was circulated to the potential specifiers in the office for a week and then placed in the library for filing. Of the six, Products 5, 7 and 8 were never considered for specification during the monitoring period despite the fact that projects were being designed and detailed where similar products were specified. Information about these products was still in the office library at the end of the experiment, presumably by then out of date.

Comparison with the postal questionnaire

A question in the postal questionnaire was designed to measure the awareness of the same nine products by specifiers in other architect's offices in the same region. The questionnaire was posted 25 weeks into the monitoring period, and replies were received between 4 and 8 weeks later. For the purposes of making a comparison with the diary of adoption, it was assumed that they were completed during Week 28. The question asked was: 'The following products have been launched onto the market within the past 12 months, would you please indicate those you are aware of, those you have considered using and those actually specified.' The replies are shown as the number of respondents from a total of 138. The figures are compared with the diary of adoption, where the specifier's awareness/use is indicated (see Table 9.1).

Table 9.1. Comparison of diary office and postal questionnaire respondents

Product	A	C	S	Diary office
Product 1	69	15	5	Specified
Product 2	66	17	3	Specified
Product 3	25	8	2	Unaware
Product 4	31	3	2	Aware
Product 5	37	2	4	Aware
Product 6	19	0	0	Unaware
Product 7	48	6	1	Aware
Product 8	24	13	1	Aware
Product 9	18	4	0	Unaware

Twenty of the 138 respondents indicated that they were unaware of any of the products. Key: A = awareness; C = considered; S = specified.

Fifteen weeks into the observation, Product 1 had been specified by the office and recorded both the highest awareness and highest level of specification by respondents to the postal questionnaire. Product 2 had also been specified by the diary office, recorded the second highest awareness in the questionnaire, had been considered by 15 respondents, and had been specified by three of them. Of the four building-product innovations with the lowest awareness in the postal questionnaire, three of these had *not* made it past the two gatekeepers in the diary office. This may be coincidental, it may be a reflection of the marketing strategy employed by the company, or they may have met with similar resistance in other offices. Nevertheless, the similarities between the postal questionnaire and the architect's office in which the monitoring was taking place suggest that the actions monitored were representative of behaviour in other offices.

The specification of 'new' building products

The events recorded during the observation period are presented here under the product number.

Product 1

This manufacturer's trade representative had been invited into the office to discuss the possibility of using their bricks for a particular project. During the meeting (with one of the authors) in Week 1, the representative took the opportunity to introduce the coping system (Product 1) and left two copies of literature; one for the office library and one for personal use (i.e. for inclusion in a palette of products). The information was placed in the technical partner's 'in tray'. The introduction of this product was very timely, since it appeared to solve problems that the practice were having with vandalism to boundary walls (traditional detail) on several inner city sites. A week after his visit, the trade representative delivered a sample of the product (a 'tactile demonstrator') to the office, and this too was placed on the technical partner's desk. It also attracted considerable interest from the specifiers in the office. This well-known manufacturer believes that if a specifier can actually handle a product, the chances of specification are much higher because the product can be explored in greater detail than paper information allows. Their trade representative also offered to take specifiers to their brickworks to see a demonstration of the product being built into a wall. This was declined because the tactile demonstrator served a similar function.

One of the specifiers specified Product 1 as a trial on a boundary wall for a new project during Week 3. His decision to adopt this product as a substitute for a traditional tile creasing detail was made after checking technical queries with the manufacturer's technical department by telephone. Approval was also required, and given, by the technical partner, before seeking approval from the client, which was also granted.

The product was not delivered to site as programmed because of manufacturing difficulties at this early stage in the development of the product. Most of the innovations being considered here are only innovations in that they were new to particular specifiers. This product was also an innovation in the commonly used sense of the word, i.e.

it was a new product. Teething problems with new products are an additional factor that, in some cases, can lead to rejection or discontinuance of the product (it has not been possible to explore this effect). The product was eventually delivered to site 2 weeks late. Although this led to a request from the contractor to change to a traditional detail, to save time, this was declined by the specifier. This is an example of pressure-to-change products because of time pressures associated with the building programme.

The client for this building made frequent visits to site and, having seen the coping system, said that he liked it and asked the technical partner to use it on all future projects. Standard detail drawings for this particular client were updated to include Product 1 on boundary wall details, thus ensuring automatic specification on all future jobs. However, standard details for other clients in the office were not revised because the technical partner did not feel it necessary. The adoption of some innovations may be client-specific, suggesting that offices may have other files of approved products for repeat use with regular clients that may differ from the normal office palette.

During Week 51, the technical partner was telephoned by a specifier working in an architect's office elsewhere in the country who also worked for the same client. They had been asked by the client to specify Product 1 on all of their future projects and had telephoned to seek further advice (and peer group approval), since the product was new to them: this led to the other office adopting this product as a standard detail. The two architectural offices had an informal relationship because they shared the same client, which meant that technical knowledge was occasionally shared.

Evidence of potential discontinuance came during Week 55 when the quantity surveyor suggested that the specifier should revert to a traditional boundary wall detail to save money (on a project that was running over budget). By this time, the product had been specified on seven projects because it was a standard detail. The request was declined, but the issue of cost was picked up by the client, who then asked the practice to only use the product on future projects where absolutely necessary. Thus, it was only used as a standard detail where vandalism was thought to be a problem. During the 40-month period, the building product innovation was adopted quickly on projects for a particular client, but not for other projects. Pressure to change came from the quantity surveyor initially, then from the client.

Product 2

During Week 8 of the monitoring period, one of the specifiers had to re-design a building to suit the request of a planning officer, who had refused to accept the proposed cladding material, timber boarding, to the gable walls of a building. The material had been annotated on the drawings submitted for full planning approval, and the planning officer had stated that he would recommend that the application be refused unless the material was changed to 'something better'. There was evidence here of contribution to the specification process from someone outside the social system, in this case exerting pressure to change the material already selected by the specifier.

It was standard office policy to report any changes in design and the reasons for them to the client. In this case, the client (the same as above) asked the specifier to resist the requested change. This resulted in a number of telephone conversations and an unsucessful meeting with the planning officer. At the meeting, the planning officer said that he would accept brickwork or render. Brickwork was rejected by the specifier because it would have required changes to the design of the structure of the building but, more importantly, would have required the revision of a number of drawings and the preparation of two new detail drawings. The use of render was also rejected by the specifier because of the anticipated maintenance costs of the material. A material was needed that could be substituted for the timber boarding without the need to alter too many drawings.

Information about such a material, Product 2, had recently been circulated to specifiers in the office and placed in the office library, and the specifier went there to look for suitable products and took out this information. Information about similar products was also available in the library, but was ignored by the specifier because Product 2 was fresh in his mind (he had remembered its being circulated in the office). The information about similar products had been in the library for some time, which may have influenced the decision to ignore it. This incident highlights the value to the manufacturer of a specifier becoming aware of an innovation during or just prior to his need for it. Of course, had this information not been so recently circulated, the specifier may have paid more attention to the alternatives.

He telephoned the company to request samples of the product and for a trade representative to visit the office to discuss its use in more detail, that is he was seeking reinforcement. The representative visited the office the following week and satisfied the specifier's questions about the cutting and the fixing of the product on site. Guarantees and a technical specification were provided by the company and checked with the technical partner, who approved its use and who also obtained the client's approval. Drawings indicating the use of Product 2 and a sample of the product were taken to a meeting with the planner the next week and were approved: the building product innovation had been adopted by the office over a 3-week period. The specifier placed a separate copy of the trade literature (which had been brought to the office by the trade representative) in his personal collection and, when questioned about this, he said it was so that he could refer to it easily if there were any queries from the building site, rather than having to get the information out of the library. He also liked it and intended to use it on future projects.

During Week 33, the product was delivered to site, and, following a site visit, the specifier was overheard telling a colleague in the office that it was a good product. However, after the product had been fixed on site, the client said that he did not like the finished appearance. He preferred the timber boarding and said that this product should not be used again on any of his buildings. Although the building product had been adopted by the specifier, there was no opportunity to specify it again until Week 149 (no one else in the office had specified it either). It was for the same client, and once again, a planning officer in a different planning authority said that timber would not be approved. The client stated quite strongly that '... on no account was the practice to use this product.' There was nothing wrong technically with it: he merely did not like its appearance. As a consequence, it was not specified, and a traditional render detail was used (despite the concern over maintenance costs). Despite this, the specifier retained information about this product in his palette.

Product 3

This was the product that the technical partner had initially rejected because he perceived it as too risky to use. Following his initial rejection, requests were made by a number of building contractors to the technical partner for the office to specify a simpler and quicker system of closing the cavities at openings in the wall. Both Product 3 and a competing system were suggested by three different contractors over a 12-month period. In this case, the product had already been used by the building contractors (specified by other architectural offices) who found the product both easy and quick to fix on site. The contractors had found that the product had a high relative advantage, was compatible with existing construction details and was simple to understand. Another important point was that they had been able to use the product on site without risk to themselves (because it had been specified by an architectural office who therefore carried responsibility for its selection). At this time, two of the contractors had adopted the product as a standard detail for projects on which they were responsible for specification (i.e. design and build projects and speculative housing developments). It was the contractor's experience of using the product that was influential in the technical partner's decision to re-assess the information.

In this example, knowledge about the innovation was communicated to the potential specifier by a source other than the manufacturer, through interpersonal communication between specifier and contractor, which was also influential in reducing the technical partner's concern about the innovation. The product innovation had two advantages: it offered improved thermal insulation (improved performance), but was initially rejected on the grounds of offering insufficient advantage verses risk, and it offered ease of installation on the building site (improved buildability), and was eventually accepted by the specifier at the request of the contractor. The specifier was concerned with the first characteristic, whilst the builder was concerned with the second.

Technical information and a sample of the product were requested by telephone directly from the manufacture and sent to the office by post. At that time, the product had been on the market for about 2 years, and the company offered names of other architectural offices who had used it in an attempt to endorse the product. Information about the other, competing, building product innovation suggested by the contractors was ignored by the technical partner, simply because he had seen both products on site and favoured Product 3.

The technical partner checked the product's technical specification, from the literature, and then asked one of the specifiers in the office to revise the working drawings for one project, which was carried out during Week 70. Thus, the product innovation had been adopted on a trial basis: none of the other jobs in the office was altered at this time. The product was fixed on site during Weeks 96–98. There were no problems with either delivery or fixing, and during Week 118, the technical partner issued an A4-sized drawing, together with an internal memorandum, which asked all staff to revise the detail on all future jobs to include Product 3. It had been adopted by everyone in the office through what Rogers describes as an authority decision. At the end of the monitoring period, the product was still specified on the standard construction details for all new build work.

Product 4

Information about Product 4 was circulated in the office and then placed in the library. No interest in the product was recorded until Week 68 when a structural engineer suggested its use during a telephone conversation with a specifier, in which they were trying to find a solution to a design problem. A photocopy of the technical information was sent from the structural engineer to the specifier that was perceived as an innovation (he had forgotten that he had seen the literature earlier). In this example, the structural engineer contributed to the specification process by providing knowledge about the product to the specifier: awareness was not directly from the manufacturer but through a third party, an external contributor.

Further technical information was requested by the specifier (it was already in the library, but he did not go and look), and a trade representative delivered this and a sample of the product to the office during Week 71. A decision was taken to use this: the structural engineers had used it previously when working with other architects' offices, and information was sent to the quantity surveyor for cost appraisal. At Week 74, the quantity surveyor reported that its inclusion would put the estimated cost of the project over budget. After discussion, a revised scheme was developed by the structural engineer and the specifier that did not require the use of this product. Although it was not adopted in this case, information about it was added to the specifier's palette of favourite products for future use (postponed adoption). It was not used during the monitoring period and was still in his collection of personal literature at the end of the experiment, suggesting that he may use it if a suitable situation was to occur.

Contributing factors

During the monitoring period, it became apparent that a number of contributing factors were present that were not related to any specific building-product innovation but were important in relation to the application of the Rogers model to the building industry, and it is important to note these. Some general observations can be made about the behaviour of specifiers in the office, and there was some influence on the events because of the movement of staff that occurs between offices. The contribution of Local Authority town planning officers also requires some discussion.

The specifiers' behaviour

During the experiment, the practice received certification for its Quality Assurance scheme. This should have affected the specifiers in the office because there was a clause that theoretically stopped individual members of staff retaining their own source of literature. This was to prevent the retention of out-of-date material, and, according to the quality manual, they could only use information from the office library. In fact, what happened was that individual members of staff simply kept their personal collections of literature away from public view, taking them off open shelves and locking them in the bottom draw of their desks. By this means, they continued to use them, so continuing the tendency to use the individual palette before looking elsewhere. One might well ask whether strict enforcement of the quality manual would have brought specifiers into contact with a greater range of trade literature and increased their knowledge base and so possibly the number of innovations considered.

Staff movement

During the monitoring period, there was an economic recession that seriously affected the building industry. This meant that there was a tendency for staff to stay in their current employment (unless made redundant), rather than to move jobs frequently, as had happened in the economic boom of the late 1980s. The office in which the data collection was carried out, unlike the majority of other architect's offices at the time, did not make anyone redundant because of the recession. This meant that the office was a relatively stable environment in which to carry out the experiments, although three of the specifiers did leave the office to take up employment elsewhere and were replaced with three new staff with similar qualifications. This provided an opportunity to monitor any information transfer between offices through the movement of staff.

The first to leave (Week 63) was replaced by someone of a similar age and experience who had been working in a smaller architect's office in a nearby town. The outgoing member of staff took his personal collection of literature with him, while the new member of staff brought his to the office (despite the restrictions described above). The specifier who had left the practice was interviewed 12 months after his move in an attempt to see if he had introduced any product innovations to his new office. His new office had adopted Products 1 and 2 as approved products before he took up his new job, and he had specified both (although he had not used Product 1 while in the office being monitored). He was unable to confirm whether or not he had introduced any new products to his new office during the interview. The other two who left took up jobs that took them away from product specification.

There was no attempt to analyse the content of specifiers' personal collections of literature and there was no evidence that the three new members of staff introduced any building product innovations to the office being monitored. All three new members, did however, specify Products 1 and 2 because they had been adopted by the office and were included in standard details and the master specification for certain clients.

Planners' contribution

During the experiment, there was resistance to the use of brick on a building project. When the planning drawings were submitted for approval, the planning officer asked the specifier to change to artificial stone. Following a number of lengthy discussions with the planning officer, she finally granted approval for the use of brick. In this instance, the pressure from a planning officer exerted on the specifier to change materials would have led indirectly to the rejection of Product 1 on this particular project; however, two other factors emerged.

The artificial stone manufacturer's trade representative had spent a lot of time convincing the planning officers that his company's product was worthy of consideration; he had seen all of the planning officers in areas where stone had commonly been used in the past, and had placed trade information and product samples in the office library of the planning departments. Clearly, the representative had done an excellent job because the planning department were legally outside their remit in insisting on the use of this company's artificial stone products: they could recommend the use of a stone treatment but not a particular company's product. But significant here is that awareness of a building product had come from the planning officer, an external contributor.

Secondly, when the town planner was interviewed, it was found that in addition to the planning department's extensive library of trade literature and samples, mainly external materials, the planning officer had her own personal collection of favourite materials. Clearly, the manufacturers and their trade representatives were aware of the planners' role in product selection and saw it as another route to raising specifiers' awareness, albeit indirectly.

Implications

Because the monitoring took place over a long period of time, a number of events were recorded that might have been missed in a shorter period, such as examples of discontinuance (Product 1) and adoption after initial rejection (Product 3). The diary of adoption highlighted the complexity of the specification process, in particular, the contribution made by persons from outside the architect's office. This demonstrated how information about building-product innovations can enter the architect's office through interpersonal communication channels generated by the building project itself, rather than directly from the manufacturer, a process that bypasses any formal gatekeeping mechanisms.

The filtering of trade literature as it entered the office was formally controlled by two gatekeepers, and evidence of this supported the gatekeeping research reported earlier. However, despite the existence of a quality management system that had regulations supposedly ensuring that all trade literature in the office was up to date, the specifiers kept their own collections out of sight of the office quality manager. Some of this had been brought from other offices, and some of it must have been out of date.

The specifier's palette of products was used to store information about products that had been adopted and also to store information about products that had been investigated but for whatever reason were not specified. This was so that they might be considered for use in the future, a new condition referred to as postponed adoption. Furthermore, there was evidence that the external contributors, certainly the planning officers, also kept a palette of favourite products.

Awareness of new products was clearly linked to need. When a specifier had a need, the building-product innovation was adopted quickly, as was the case with Products 1 and 2. Where there was no immediate need, there was a tendency to ignore information about the innovation or forget about it (e.g. Product 4).

Two of the building-product innovations, Product 1 and Product 3, were both made of uPVC, a material disliked by the technical partner and both used in conjunction with brickwork. So why did he initially adopt one and reject the other? Product 1 was manufactured by a company known to the office and hence was viewed as carrying little risk. Product 3 was a new name to the office and was perceived as carrying a greater risk because the manufacturer was unfamiliar. This shows that the track record of the company promoting the building product innovation was an important factor at the awareness stage for this individual, an observation that contradicts the views recorded in the postal questionnaire. There was no evidence that the cost of a particular product was considered by the specifier. This was left to the quantity surveyor and only addressed by the specifier when identified as a problem.

So, the manner in which the specifier's office is managed, the personality of the individual specifiers, as well as the influence of others party to the construction process will influence the uptake of new products. So, too, will the characteristics of individual products.

Specifier observed

This chapter is concerned with an individual specifier's decision-making process and the pressures faced to change product specifications during the design and assembly phases of a project. The case study is presented as a series of steps from which the implications are discussed with relevance to best practice.

Observing detail design decisions

We have seen some discrepancies between the findings of the questionnaire and those of the diary of adoption. In some respects, the way in which people believe that they behave does not seem to be the way in which they actually behave in practice, and this has been shown by the closer focus provided by the diary. The interviews that were carried out also showed the varying effects of office policy, in particular how difficult it might be for some products to get past the initial gatekeeping procedures. But it also showed mechanisms by which these might be bypassed. It showed that the adoption or rejection of building products might depend upon the very particular circumstances of individual jobs. That being so, it is going to depend upon the behaviour of the individuals involved in the process.

In circumstances where the specifier is unable to use his or her palette of known products and seeks alternatives, it is sometimes a matter of choice between competing innovations. This is not something that has been considered by the Rogers model but which is useful to explore to see how the basic model of diffusion needs to be adapted for the design situation. The question is, how does the behaviour of the various players affect the adoption or otherwise of an innovation? This chapter therefore focuses on an individual specifier's innovation-decision process using a case-study approach. The case study is concerned with the factors that cause a specifier to look beyond his palette to building-product innovations and the ensuing innovation-decision process.

The observation is described as a series of distinct steps, which need to be incorporated into a theoretical model.

The specifier

At the heart of diffusion research is the 'innovation-decision process' that describes five stages through which a potential adopter may pass (Rogers, 1995) (see Figure 7.1) and which have parallels with the specification process. The specifier will pass from first exposure to information about the innovation (knowledge), through a period of gathering more information to consider its characteristics (persuasion), to making a decision to use or reject the innovation (decision), to construction on site (implementation) and the intention to use the product again (confirmation). The innovation-decision-making process is important because the individual selects a *new* product over that previously in existence. Thus, the *newness* of the alternative is an important aspect of the innovation-decision-making process, described by Rogers (1995:161).

This process is going on when the specifier is detailing the building; as such, it will be influenced both by the particular design project being worked on and by the amount of time available to the specifier: it is not carried out in isolation but as part of the detail design decision-making process. As noted earlier, the design of a building will involve a matrix of decision-making that will vary in complexity as the design progresses. A number of specific decisions are taken at each stage that in turn influence or determine what succeeds them. Although the RIBA Plan of Work indicates a theoretical framework for decision making, in practice the sequence of decision making is not always adhered to, consisting of a number of skipped stages and constant re-assessment. Each specifier will have his or her own subjective perception of the problem, based on past experience and possibly the past experience of the office (i.e. previous jobs of a similar nature). He/she will attempt to find the action that will be *satisfactory* and not necessarily *optimal* in meeting his own objectives. As discussed above, the decision-making will be influenced by the individual's personal characteristics (status, age, personal values, etc.), the situation (such as the type of building or the stage of the project) and the amount of time available in which to complete the process (degree of urgency). Furthermore, the contribution from persons outside the social system cannot be ignored.

Before this process starts, however, the specifier must have identified a problem that cannot be resolved from the information contained in the collection of favourite products, hence triggering a search for information and the start of the innovation-decision process. The challenge was to try and observe this process.

Methodology

Two approaches were considered, setting up an experiment and direct observation in an architect's office. The first was discounted because it was prone to the same problem as the diary method, i.e. raising the specifier's awareness to the research. Direct non-intrusive observation of a specifier appeared to be a more natural approach, but one with its own methodological difficulties (Nason and Golding, 1998), discussed below. Observation had the potential to identify and illustrate the tasks involved from within the organisation. Consistent with ethnographic research, the goal was to interpret the behaviours of the social system being studied (Rosen, 1991), an approach adopted successfully by Dana Cuff (1991). Because the author was working in the design office of an established architectural practice, the opportunity to observe specifiers in action was available.

The objective was to observe a situation or event that caused a specifier to investigate 'new' products; therefore, the specifier(s) would be self-selecting and the period of observation determined by the individual and the project life cycle. There was a danger that the critical event may not have occurred when the observer was present. What was significant was that the opportunity for observation did not occur for a considerable amount of time because specifiers in the office continued to select products that were familiar to them, thus demonstrating that the adoption of building product innovations is a rare event, supporting the views of Mackinder's sample. Eventually, a situation arose that could be observed and recorded for the duration of the process. The specifier observed sat next to the author and had the unusual habit of talking out loud whilst he was working: although this irritated other members of the office, it made for an ideal subject since the thinking process and the decision-making process, usually hidden from an observer, were quite transparent. The observations were recorded by the author in writing in a desk diary, recording both the actions observed and the length of time taken, and then analysed. At the end of the observation process the specifier was told that he had been observed, and his consent was obtained both to use the material gathered and to interview him to explore the motives behind his decisions.

As with all ethnographic research, the material gathered was a detailed account of a specifier and the influence of his direct surroundings. The specification act was highly interrelated to other activities within the office, and the process reported below had to be separated out from other tasks that engaged the specifier during his working day (working on two other projects concurrently, attending meetings, dealing with telephone calls, site visits, etc.). The significance of the findings to this particular design office cannot be ignored, but the observation reported below helps to illustrate the complexity of the decision-making process that occurs during the specification process, as illustrated in Figure 10.1.

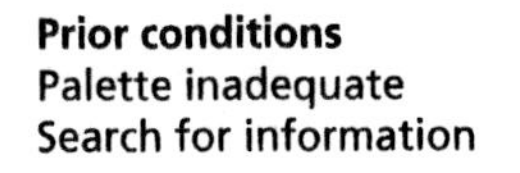

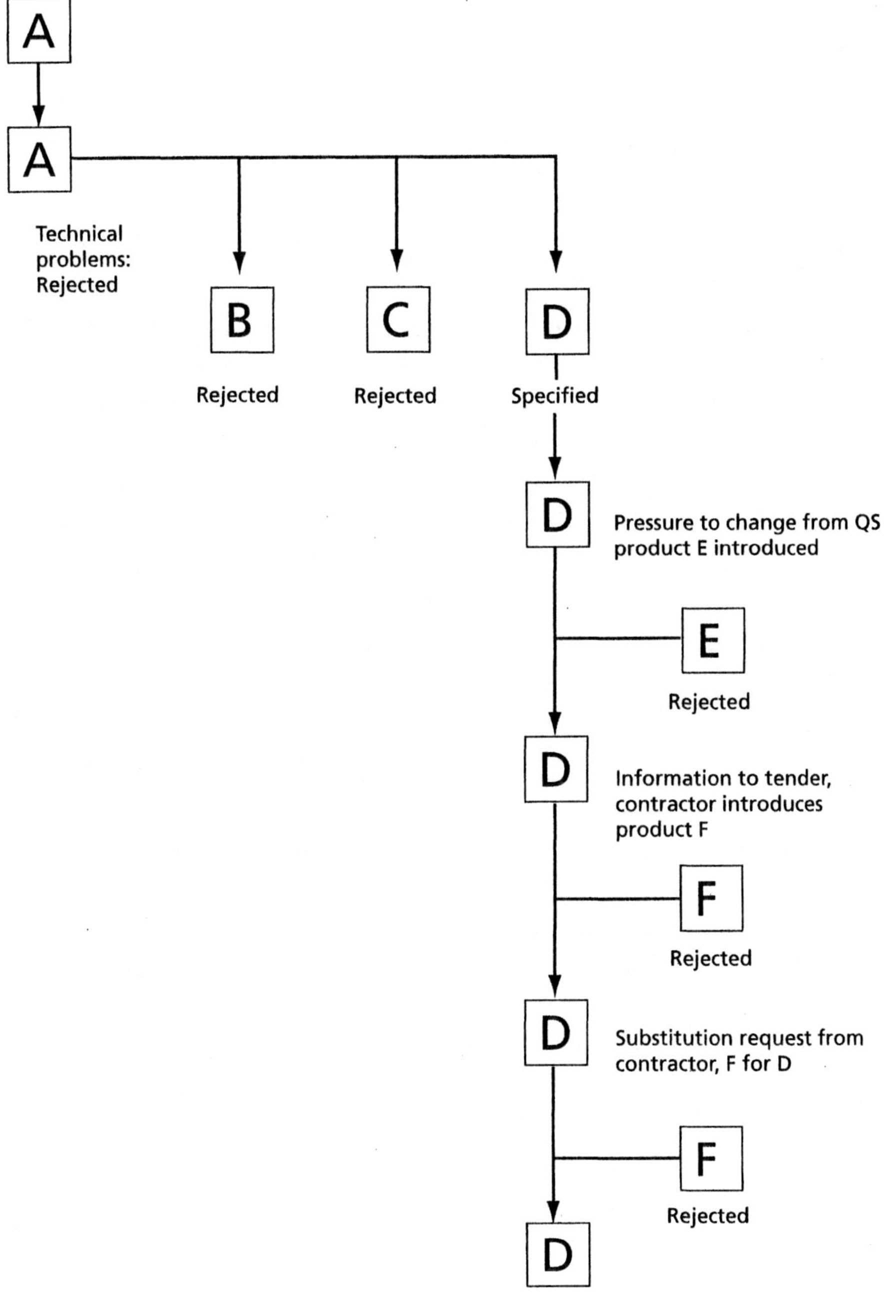

Fig. 10.1 Steps in the specification process

A search for 'new' products

The specifier had 2 weeks to produce the working drawings for four, single-storey, retail units. Three of the four were to be built with timber rafters and concrete interlocking roof tiles, and the fourth was to be detailed with a structural metal tray and a profiled steel roof to suit a particular client's requirements. He first detailed the three units with the tiled roof, a task carried out quickly because he was familiar with this form of construction. He had used a very similar roof construction on a previous project, and the drawings produced for it were used to gain information for use here, thus reinforcing the tendency to use familiar products, observed in earlier research (Mackinder, 1980). When he attempted to detail the metal roof, a form of construction that was unfamiliar to him, he was unable to draw on his previous experience because he had not worked on any similar projects, although other specifiers in the office had. Since his personal collection of literature did not contain any information that could help him, he was forced to look for products that may solve his particular need: he was forced to search for information about building products that would be new to him, (consistent with Stage 1 of Rogers' model), i.e. he started a different pattern of behaviour.

Innovation A

His first action was to ask other specifiers in the office if they had experience of detailing such a roof, so he first sought knowledge from his peers, drawing on the collective experience of the office, an action he later said was taken to 'save time looking in the library'. A colleague suggested a product that the office had used successfully before, Product A, but which was new to the specifier. He spent approximately 10 minutes talking to his colleague to gain more information and to establish whether or not the product was suitable for his particular requirements. He was seeking to reduce his level of uncertainty, which is consistent with the persuasion stage of the model (stage 2).

Satisfied with the information gathered, he then sought further information about the product from the office library to enable him to make a decision. Because the trade literature was not comprehensive enough to solve all of his queries, he telephoned the manufacturer to request additional literature. The manufacturer offered to send a trade representative, the change agent in Rogers's terms, to the office to assist with any queries: this was declined by the specifier, who later said that he did not have sufficient time to see the representative. Information was received by post 3 days after the request (during which time the specifier had been working on another project). After reading the information, he made a decision to specify Product A and continued with his detail design work. The innovation had been adopted, consistent with stage 3 of the model 'decision', but his decision was subsequently revised.

While detailing the roof, he discovered a technical problem that he could not resolve from the literature, so he telephoned the manufacturer's technical department for clarification. During the telephone conversation, it became clear that Product A would have to be modified to resolve his particular problem, but the manufacturer did not have a 'standard solution' simply because they had not considered such a possibility arising. This resulted in a state of dissatisfaction on behalf of the specifier, who immediately went to the office library to search out an alternative. He did not use the electronic database or the printed product compendia: he selected entirely from the trade literature on the library shelf, a search pattern that the specifier later confirmed to be the quickest way of finding suitable products, further emphasising the time pressures exerted during the detail design stage.

Innovations B, C and D

From his search in the library, a further three building product innovations were selected, a task that he only spent 10 minutes on. The library contained trade literature from 10 manufacturers of similar metal roofing products, seven of which were rejected simply because their technical details in the literature was seen to be of 'poor quality' by the specifier. On returning to his workstation, the specifier telephoned all three manufacturers to question them about their products. Product B was discounted because the technical representative was perceived as not knowing his product well enough (described as a 'complete idiot' by the specifier). Product C was rejected because the company would only answer technical queries by sending a trade representative to the office; since the earliest appointment would be too late for the specifier to complete his task to programme, the product was rejected. Product D was adopted because the technical representative 'knew his stuff' and had offered some additional practical advice to the specifier that helped him to complete his detailing quickly. In line with office policy, the specifier went and spoke to the organisation's technical partner, who was responsible for granting approval for the use of any product that was new to the office. Following a short discussion, approval was granted, and the specifier returned to his desk to resume his detailing of the roof. Product D was referred to by name on the drawings and later in the accompanying written specification, a task he completed within the 2-week programme.

Pressure to change

The production information was then sent to the quantity surveyor for production of the bills of quantities and also for a cost check of the design against the original budget. During the 3-week period taken to complete this task, the quantity surveyor telephoned the specifier to suggest an alternative to that specified in order to save money. The alternative, Product E, was unknown to the specifier, so it constituted a further building product innovation, one introduced to him by a contributor to the design process who was primarily concerned with the cost of the product, not its technical performance. This illustrated the contribution made from outside the architect's office during the specification process, with pressure to change the specified product and also the introduction of a product that was known to the quantity surveyor but not the architectural office.

Product E was immediately rejected by the specifier simply because he had invested a lot of time in solving a particular problem and did not want to go through the process again with a different product and different fixing details. He made no attempt to analyse the information, despite the potential cost savings reported by the quantity surveyor. As a result, Product D survived this first attempt at specification substitution and was included in the documentation sent out to competitive tender. This illustrated the contribution made from outside the architect's office during the specification process, with pressure to change the specified product and also the introduction of knowledge about additional product innovations, confirming one of the stages of the theoretical model, and also again illustrates the effect of time on the process. This time constraint is specific to the design process and not part of the Rogers model. Time pressures appeared to be of paramount importance to the specifier.

Evidence of specification substitution

The lowest tenderer was accepted by the architect's office and approved by the client. But the contractor had also submitted an alternative, lower, contract sum based on a revised specification. Twenty-three products had been identified for which there were cheaper alternatives ranging from the facing bricks and cavity insulation to the ironmongery for the internal doors and including one to replace the steel roofing system, Product D. Thus, a further innovation had been introduced, Product F, on the contractor's list of suggested alternatives. Individual product costs were not listed; instead a total cost saving had been identified, compared with the original products specified. The client asked the specifier to analyse the alternative tender sum and then advise him which to accept. The proposed alternative, Product F, was unknown to both the specifier and the office.

The client asked the specifier to analyse the cheaper specification. Although the specifier wanted to reject the substituted products immediately, further information had to be sought so that a report could be made to the client. He telephoned the manufacturers and asked a number of questions about delivery and guarantees. The answers raised further issues to be investigated, and since he did not have the time to pursue them, he rejected all of the substituted products, including Product F, recommending to the client that the cheaper products were of insufficient quality and/or not acceptable visually (e.g. the facing bricks). This whole process, including the writing and faxing of the report to the client for approval, was dealt with in 24 hours. The contractor was appointed the following day with no change to the original contract documents. Product D had survived.

Further attempts to change a number of products, including Product D, were made by the contractor after the project had started on site on a number of occasions during the 30-week programme. This confirmed evidence of specification substitution reported in earlier work, although in the event, the specifier refused all requests. First, the contractor claimed that the specified product could not be delivered to suit his programme and proposed Product F again. This was found to be untrue when the specifier checked with the manufacturer, who confirmed that the contractor had made no attempt to place an order for Product D (presumably, the contractor had hoped that the specifier would accede to his wishes without checking). The request was refused. After the first request had failed, the contractor again proposed that Product F be substituted to save money (for whom was never made clear). Again, this was refused by a specifier keen to see his design decision transferred from drawing board to finished building. Eventually, Product D was delivered to site, to programme, and built into the building without any problems being reported from site. Thus, after a number of attempts to change it, it had finally been implemented (stage 4 of the model).

Toward the end of the project, the specifier added Product D to his personal collection of literature for use at a future date. It had now become part of his personal inventory of products. This could be seen as evidence of the confirmation stage (stage 5) because the likelihood of the product being used again is high, although not observed in this study.

The persistence of the contractor deserves some comment here. Although no enquiries were made to explore the reasons for this, there are a number of clear possibilities. The most likely is that Product F was one that the contractor had used before and therefore presented no problems for him. If he was unfamiliar with the specified product, he might be uncertain about his ability to use it without difficulty. Of course, it is also possible that he was only too familiar with the specified product and had previously experienced some difficulty in using it. Another possibility is that Product F was available through the contractor's normal supplier and the amount of discount greater than that for the product that was not so available – thus, financial considerations affecting the behaviour of the contractor cannot be ruled out.

Talking about the act

When the observation period was complete, the specifier was interviewed to address questions about the process of specification and his attitude towards building-product innovations. He was asked to recount the actions he went through. Although the specifier was interviewed immediately after the project finished, he was unable to recount all of his actions, providing a rather generalised account of events, failing to describe the dead-ends and unable to remember how many attempts were made to change the product, consistent with Yeoman's (1982) findings. While this helped to justify the ethnographic approach adopted, it meant that the interview had to be adjusted to gather the specifier's attitudes towards product selection rather than as a tool to expand upon his observed behaviour.

Although the specifier described himself as creative and always looking out for new products he was aware that his actual behaviour was contrary to this. He claimed that he was 'forced to be conservative' about product selection and detailing because of his, and his organisation's, concerns about building-product failure. Products that were new to the office carried a perceived enhancement of risk. His risk-management technique relied on the specification of products that he had used previously or, failing that, those used by the office. His collection of literature had been assembled over a long period in the building industry (25 years) from products that he said were 'known to perform', i.e. he was pretty confident that detailed and implemented correctly, these products would not fail.

Information about a building product innovation would be added to his palette if a new situation had required its consideration and if there were no problems in specifying it and no problems reported from site during or after construction. Trade representatives were only seen or spoken to by telephone when further information was required for a specific project. Thus, communication with building-product manufacturers was always initiated by the specifier, which confirms the view that specifiers have clear active and passive modes.

He also said that he tried to stick to products that he had used previously because the time pressures imposed on him by both the design programme and the construction programme rarely allowed him any time to investigate alternatives. At the time of the observation, the specifier was working on three other jobs, all at different stages, and all with demanding programmes. Time constraints are not included in the Rogers model because the situations that he has discussed are not affected by these. However, they are clearly important in the design process. If the time from awareness of the innovation to obtaining adequate information to enable a decision to adopt is too long, then the innovation will be rejected.

From this interview, two reasons were noted for the specifier to look outside his palette: technical substitution and a novel design problem. Technical substitution would occur if the product in the specifier's palette was unsuitable for the given situation. This would result in a search for information about other products, which may themselves eventually enter the palette. A novel design problem might, for example, occur if the specifier was engaged on a different building type from that normally commissioned, resulting in the need to search for different types of building products. Both of these situations would result in the specifier engaging in an active search for information.

There was no opportunity for the specifier to use this product again during the period when his actions were being monitored, so the confirmation stage could not be observed. However, when asked if he would use the product again, he replied, 'Yes, if I need to'.

Reflection on the observation

Before any conclusions can be drawn, it is necessary to comment on the method used. Because the author was responsible for the day-to-day management of the design office in which this individual worked, it is possible that the observer influenced the behaviour observed. However, throughout the observation period, the specifier did not seek any clarification of his decision-making process from the author; rather, he sought approval from the technical partner in accordance with office policy. There is also the possibility that the observer missed part of the process, but retrospective analysis of the written evidence produced by the specifier, namely the drawings, written specification and notes in his desk diary, supported the observations.

Ethnographic research produces unique findings that are difficult to generalise from. The action reported above was influenced by the specifier's office environment, time pressures, characteristics of the project and the characteristics of the specifier. Naturally, the question has to be asked as to how representative this behaviour is of other specifiers. When talking about his behaviour, his opinions were consistent with Mackinder's sample of architects, despite the long time gap between the two pieces of work. His behaviour was also consistent with that of other specifiers reported in Chapter 9. There was no evidence to suggest that the actions recorded were unrepresentative.

This specifier and other specifiers in this office used the prescriptive method of specifying, specifying products by proprietary (brand) name, and this observation has identified pressures to change associated with this method of specifying. In organisations where performance specifying is used, the final choice of proprietary product rests with the contractor, and thus, the process and pressures to change will be different to that reported here. Clearly, specifiers working in other design offices will do things slightly differently, and there is a need for further naturalistic forms of enquiry to compare with the findings reported here.

Discussion

At the time of the observation, the specifier was working on three other jobs, all at different stages, and all with demanding programmes, and thus, the potential for investigating manufacturers' claims as to the performance of their products was very limited, serving to reinforce the established products. Again, there are parallels with medical research. Studies into repeat prescribing (Harris and Dajda, 1996) found that medical drugs were prescribed, without further reference to the doctor by the patient (primarily to save time), thus reinforcing the use of a familiar drug. Like the patient's drugs, the products have not been re-assessed, merely applied because they worked successfully before.

One of the issues highlighted through the observation was the impact of other parties to the specification process, a characteristic not present in the studies of repeat drug prescribing or the large body of diffusion of innovations literature. At different stages in the innovation-decision process, contributions were made from outside the architect's office by individuals with different priorities to those of the specifier. Pressure to change specifications is something that a specifier has to deal with, not just during the design phase but during the assembly process as well. In relation to Rogers' model (Figure 7.1), there would appear to be a need to add two sub-stages to accommodate the uniqueness of the specification process. Between stages 3 and 4, stage 3a would cover the pressure to change products during the tender stage. Between stages 4 and 5, stage 4a would cover the contractor's attempts to change products.

Implications

The observation reported above helps to illustrate some of the pressures and the complexity of the decision-making process that specifiers pass through. It helped to illustrate communication networks and pressures on the design process that was not evident in earlier research. The findings also suggest that building-product selection would appear to be a very personal issue for designers and, as noted above, a difficult process to observe.

Rogers (1995) questioned whether it was the need for an innovation, or the awareness of an innovation, that comes first in the innovation-decision process. The research reported here suggests that specifiers actively search out building-product innovations only when the need arises, not before. The implication here is that the adoption of 'new' products may face considerable resistance not just from the specifier, but also from the other contributors to the specification process. By gaining a fuller understanding of the individual's innovation-decision process, professional design offices may be in a better position to manage this critical aspect of building design. To do so, however, requires further ethnographic research, both to test the results presented here and to further our understanding of this little analysed aspect of design decision-making. Instead, what we can do is propose a model of the specifier's innovation decision-making process based on the research reported here. This is described in the final chapter.

Towards best practice

The final chapter brings together theoretical and practical considerations for all those concerned with specifying buildings. One of our aims in writing this book was to provide guidance based on observed behaviour. We have described how manufacturers market their products to specifiers, how offices filter this material and how specifiers specify. Here, we put forward a model of the specifier's innovation-decision process based on our research findings. This is followed by a discussion of the methods used and suggestions for future research into the specification process. We then discuss the need for comparative information and look at the establishment and maintenance of best practice before concluding with a brief discussion of future trends.

A model of the specifier's decision-making process

As demonstrated earlier, factors both internal and external to the social system might exert different pressures at different stages in the project. Thus, although the individual specifier may be content choosing from his preferred palette, building-product innovations might be forced on him or her by pressure from, for example, a planning officer or client. Therefore, building-product innovations may be introduced to the specifier from sources other than the manufacturer.

At the outset of our research, it was assumed that the adoption of building-product innovations was, to a large extent, influenced by the communication of information from the manufacturer to the specifier. It has emerged that this process is far more complex than that. Although communication of information about building-product innovations from the manufacturers is continuous, the individual specifier may employ *selective exposure*. That is, he or she will operate a personal gatekeeping system where the gate is only opened when involved in the act of building product selection. In addition to this personal gatekeeping, it has been shown, albeit from a small sample, that the partner of the architect's office exerts a considerable influence on the amount of information allowed through the office's gate to the specifiers in the office. Potential specifiers may not be aware of building product innovations, although information on them has been sent to the office because they failed to pass through the formally established gates. Therefore, they cannot be considered for adoption unless knowledge about them is gained from another source.

But, even this model is too simple, because it has been demonstrated earlier that his or her palette of favourite products would continue to be used unless the specifier is forced to look for alternatives, because it cannot meet some particular requirements. This suggests that the individual's informal gate is closed for a large proportion of the time, and only opened in specific circumstances. As important as the question posed in Chapter 8, 'How is information communicated to specifiers?', is the question 'When do designers require, or search out, information about building-product innovations?'.

As argued above, a set of conditions seem to be needed before the innovation-decision process can begin. If the palette is adequate, then the specifier will not have to spend time searching for alternatives. If, however, the palette is inadequate, i.e. cannot solve a particular problem, then the specifier will be forced to search for building-product innovations, thus starting the innovation-decision process. It would also appear from Rogers' model that this process can commence if the potential adopter becomes aware of the innovation passively. However, given the time pressures exerted on a specifier in the design office, it is unlikely that he or she will investigate information purely out of curiosity.

The model presented here is based on Rogers' model and helps to illustrate the complexity of the process. It should also help design managers to better manage the specification process through greater awareness of the issues. It should also help building-product manufacturers to reconsider their marketing strategies to specifiers (Figure 11.1).

Prior conditions

When selecting products, it seems that the specifier has (1) existing personal experience of products used previously, represented by the palette of favourite products and (2) knowledge that information about other products, building-product innovations, might exist in trade journals, product compendia or the office library. In the case study, it was the failure of the specifier's palette of favourite products that led to his search for information about building-product innovations to satisfy a particular need. So, although information about products had previously been sent to the office, via journal advertisements and listings in the product compendia, the case study showed that manufacturers depend on the particular circumstances of the design and specification

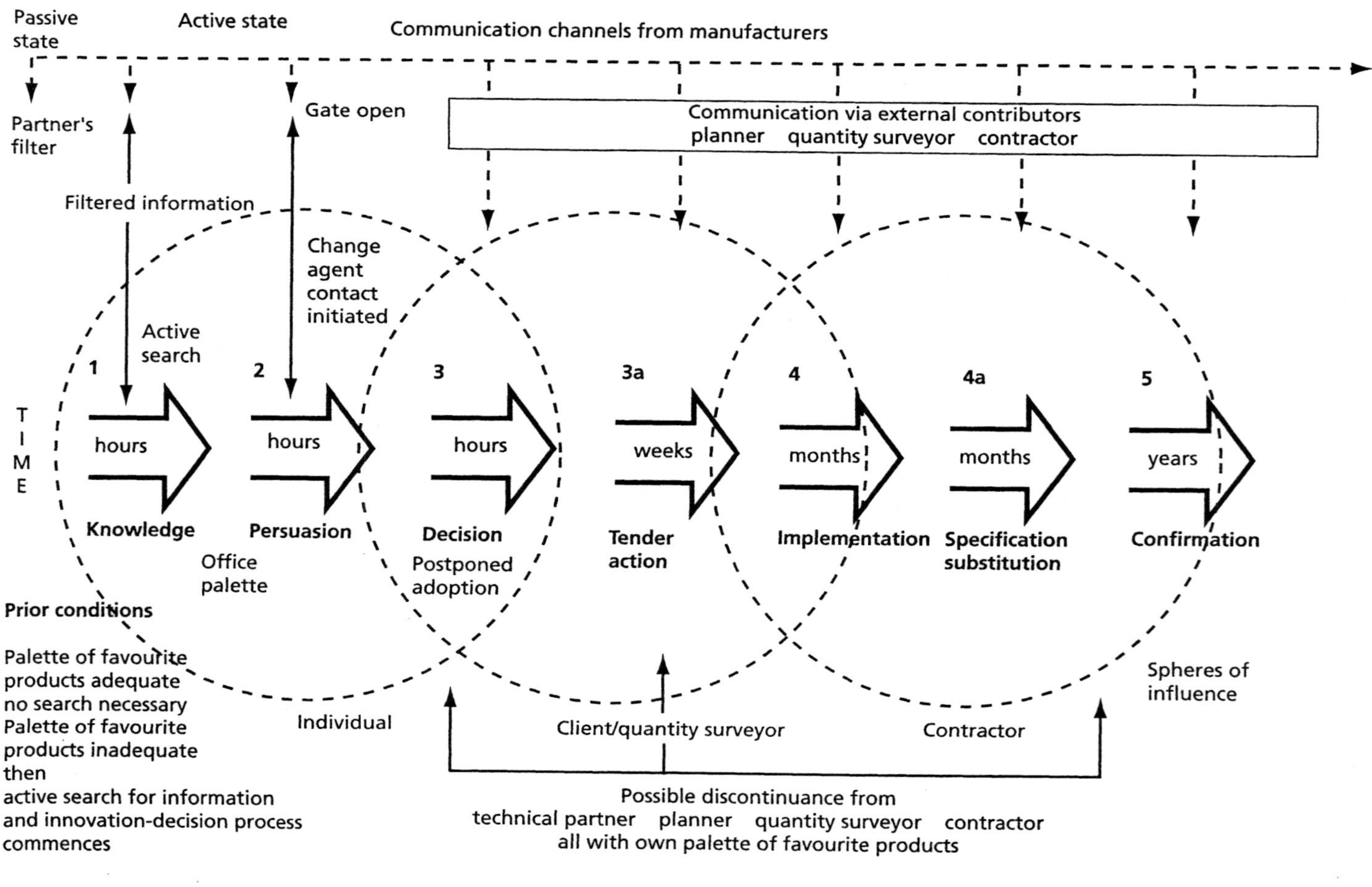

Figure 11.1 The specifier's innovation-decision process and spheres of influence

process for the specifier to become aware of a product. The specifier must be actively searching for an innovation or be made aware of it through a third party. In building design, specifiers are not passive adopters; rather they search for innovations only when the need arises. This was done through the following stages.

Step 1: Knowledge

Was sought first from colleagues in the office before looking in the library in order to save time. This requires some discussion in terms of the Rogers model because it is the office that is the unit of adoption. Available is the file of filtered, and therefore approved products, which are seen as carrying less risk than those rejected in the filtering process. The office has become aware of these products even if the individual specifier has not. There are also products that have been used previously by those in the office, which will largely be a subset of the above. When information is sought from colleagues, the tendency will be for the specifier to investigate previously used products first so that adoption of these for another project will constitute confirmation of adoption by the office.

Only if this product is not suitable in the particular circumstances will the specifier search the office library, leading to the consideration and potential adoption of an innovation. We actually have two kinds of innovation here. There are those that are innovations to a particular specifier, but not to the office as a whole, and those that are innovations to the office. The office can be said to have an awareness of them because information is within the office library and has passed through the hands of the technical partner, but they have not yet been adopted because the circumstances favouring their adoption have not yet occurred.

Step 2: Persuasion

The specifier 'tested' the company's technical department, making initial decisions based on his telephone conversation, before deciding whether or not to continue investigating the product and also whether or not to invite the trade representative into the office. This decision relied largely on the telephone conversations with the technical department during the persuasion stage. Testing each product against the particular situation might involve the making of sketch details to see if the product is suitable. This may be regarded as a trial of the product although not in the form that Rogers envisages, i.e. it is a trial in theory rather than in practice. Nevertheless, this is an important feature of the design process. At this stage, the specifier may be only partly satisfied with the product or be in a state of uncertainty because of insufficient information and therefore may invite the manufacturer's representative into the office to help and explain. Thus, the persuasion takes place in more than one stage.

Step 3: Decision

The decision to adopt the building product innovation was influenced by other parties and was far more complex than any of the examples reported by Rogers. The influence of the quantity surveyor was influential in bringing about a search for a cheaper alternative and awareness of further building-product innovations. The greater complexity of the process is a direct result of the greater complexity of the social structures compared with those examined by Rogers, and this was also seen in subsequent stages. We have also to note that if this product has already been used by the office (by a different specifier), then adoption is simply confirmation of its earlier adoption.

Step 3a: Tender action

Further building-product innovations were introduced to the specifier by the quantity surveyor and the contractor after the initial decision to adopt the innovation had been made. There were a number of opportunities at this stage for both the discontinuance of an innovation and the adoption of alternative products. Since the alternatives introduced by the quantity surveyor and the contractor were also viewed as product innovations by the specifier, this would imply that different building products can be marketed to different members of the building team.

The case study confirmed the existence of this additional stage, but the adoption of an alternative product cannot necessarily be considered as discontinuance of the innovation by the office if the product originally specified is now rejected for this particular design. Discontinuance by the office would only occur if the product now introduced were to be adopted as a permanent substitute, i.e. to be used in all subsequent designs. An aspect of building-product innovations different from the adoption of other innovations is that it does not imply total rejection of all other products. While a farmer might only sow one type of seed and might thus make a clear choice, building products always remain alternatives, each used when the circumstances are appropriate.

Step 4: Implementation

Inclusion of the building product in the documentation from which the contractor was to build the building, the written specification and drawings, is from the specifier's perspective implementation of the decision to adopt. However, this was not the end of the process because the contractor made repeated attempts to change the product during the construction phase, thus confirming the requirement for a further stage in the innovation-decision process, 4a, when considering the adoption of innovations in building.

Other professional activities may also involve decisions taking place over time and when more than one product is involved. A doctor may prescribe a new drug for one patient but may remain with a tried and tested product for another. During a course of treatment, that doctor may also switch from one drug to another for a particular patient. It is clear that the adoption process by professionals who are making decisions for a range of clients in different situations cannot be expected to show consistency of adoption. One might doubt the flexibility and thus the competence of a professional who rigidly adhered to a single product, and thus a more complex model is required for professional adoption processes.

Step 4a: Specification substitution

Despite the use of a gatekeeping system, once the project was in the contract stage, it was possible for the contractor to introduce further building-product innovations to the specifier, bypassing the gatekeepers, because they communicated directly with the specifier. At this stage, the newly introduced product was assessed against that already specified, but was resisted because of the time required to investigate its characteristics before use. However, the possibility of substitution at this stage clearly exists and is known to occur from anecdotal evidence. Therefore, adoption is only completed when the builder actually uses the product on site.

Step 5: Confirmation

It has already been noted that the adoption of products previously used by the office but new to the particular specifier constitutes confirmation by the office. Confirmation by this individual specifier was not observed, although the specifier placed the information in his palette for use in the future. Confirmation by the office might also occur through the subsequent use of the product by another specifier within the office.

The case study was concerned with product innovations that were to be used on the exterior of the building, and so they would be visible when the building was completed, which perhaps explains part of the specifier's reluctance to change his original decision. However, the specifier said that he rarely had time to investigate 'new products' because of the pressures of the building programme and noted that this aspect of his job, whilst important, had to be balanced against many other duties. This behaviour was consistent with Mackinder's sample and the respondents of the postal questionnaire who tended to act conservatively and rely on products used previously, both to save time and reduce risk.

Spheres of influence

The contribution of external agents to the innovation-decision process was far greater than had been anticipated. The main contributors were the client, the quantity surveyor, the planning officer and the contractor. Influence was exerted at different times during the process but led to both the discontinuance of adoption and the introduction of knowledge about further building-product innovations. These two kinds of influence are clearly linked since one may lead to the other, pressure to reject and adopt innovations coming from outside the specifier's office from individuals who had quite different values and priorities.

Awareness of building-product innovations from contributors other than building-product manufacturers can occur at any time after initial specification, reinforcing our contention that the process is more complex than the Rogers diffusion model. These external contributors' motives for change were in some cases driven by cost factors, a characteristic of building products that is largely unconsidered by specifiers until they are forced to do so by others. Thus, different members of the building industry are concerned with different characteristics of the innovation from those that concern the specifier.

The degree of influence from outside the specifiers' office has implications for offices because the formally established gates are bypassed. Add in the pressure to change decisions quickly, and even the best managerial systems can become ineffective control mechanisms. Clearly, for the office manager, it is important to constantly remind specifiers about the importance of their specification decisions and the likely consequences of changes suggested by parties outside of the specifiers' office.

Research matters

Before drawing some conclusions from the studies presented in Chapters 8–10, it is necessary to consider briefly the limitations inherent in the work and thus the limitations of those conclusions. The first point to note is that we have only been dealing with buildings designed by architectural practices, although there are important sectors of the building industry in which they are seldom involved. This includes contractor-led contracts with their in-house design departments and the housing sector where the kind of professional offices described here are not involved. Housing is dominated by a few large builders but with a very large number of small builders, the latter being notoriously conservative. Bowley (1960), for example, has described how difficult it was for the plasterboard manufacturers to persuade builders to use their product. At first, they both considered plasterboard inferior to the use of wet plaster and also failed to understand how it should be used. However, its eventual success depended upon its uptake by the house-building sector of the industry, and Bowley regards its widespread adoption in the 1930s as an example of prefabrication in the industry. In contrast, although a number of novel methods for prefabricated housing were introduced after the First World War, to solve the serious undercapacity of the traditional industry, this industry eventually reasserted itself, and most of the prefabricated systems were abandoned.

Nevertheless, within this conservative sector, dominated by the use of brick and block walling and timber upper floors and roofs, there have been a number of innovations that have taken hold. These have been the eventual adoption of plasterboard in the 1930s, the use of slab-on-grade ground floors replacing suspended ground floors in the late 1950s, and, more recently, the fairly widespread adoption of precast concrete flooring. Additional innovations have been the use of the TDA roof and then trussed rafter roofs in much the same period and, to some extent, the adoption of timber-frame construction in the 1970s. The adoption of the TDA roof and interlocking roof tiles at much the same time was in response to timber shortages following the Second World War (Yeomans, 1997). The return of suspended ground floors, although not necessarily of pre-cast concrete, was brought about because of the shortage of flat sites in more recent years. The impression given by these few examples is that the adoption of innovations by builders is largely in response to particular external pressures, but there is no real evidence to substantiate this as no detailed observation of builders has been made.

Within the sector of the building industry that was observed, builders were having an influence on the process. Clearly, these builders were becoming aware of new products, presumably through having to use them in response to their specification by architects. Again, their positive or negative reactions to these new products have not been observed. There were also other professionals who introduced new products to the specifiers, and we do not know by what process they became aware of them, or what factors affected their responses.

Research methodology

This research has been largely based upon the direct observation of specifiers' behaviour. Very little of the process reported was recorded on paper, and it is clear from the account given that not only would trying to trace the process after the event have been very difficult but any results would have been potentially misleading. There is always a problem about trying to trace these processes after the event because those involved are likely to remember a far more abbreviated process than that actually followed. This has already been observed in the design process when the designers' own impressions of the process after the event were compared with a more objective account based upon data collected during the event. It was then found that the designer tended to describe a logical and linear process rather than the iterative process actually followed. Dead-ends, which might well have been important as part of a learning process in the design, were simply forgotten (Yeomans, 1982). This confirms Rogers' recommendation that the adoption of innovations should be traced during the adoption process itself and not afterwards.

It had been possible to obtain a record of the design process used in the above experiment because of the rather particular circumstances of a 'team design' where thoughts had to be articulated. Note, also, that the behaviour of the specifier observed and reported here was also made possible because of rather unusual circumstances – his habit of talking to himself. The circumstances required for these observations mean that it is not a simple matter to carry out work of this kind. Researchers who wish to follow up on this kind of study might need to adopt a participant observer approach, essentially the research method adopted by workers like Cuff (1991).

Here, the methods chosen were quite different. The observations were made over an extended period of time with little direct manipulation of the process. Because normal duties in the office meant that the observer was not always present, there is the possibility that significant events might have been missed. However, there was insufficient preliminary understanding of the process to enable the design of a more controlled experiment so that this 'natural history' approach appeared to be the most satisfactory. In this case, the fact that the specifier was being observed was not apparent to him. In all of these cases, no attempt was made to change the behaviour of those being observed, nor was there any obvious temptation (or possibility) of those being observed to seek advice or comment on their actions from the observer. It is, of course, possible to envisage experiments where this is precisely what is done, i.e. where the observer is there not simply to observe but to exert some influence over the subjects of the observation in order to 'improve' their behaviour in some way.

The need for comparative information

Rogers has characterised the attributes of innovations that make them more or less likely to be adopted, and these have been discussed in relation to building products. However, the comments above suggest that a number of other attributes might be seen to be influential at different times or in different circumstances. Consider durability as an example. Under most circumstances, one would assume that this is an important attribute, as buildings tend to last for a long time. Of course, there are those buildings that are only put up for a short period of time and for which durability is not required in any of their components. There are also buildings that in themselves are durable but which contain parts that are expected to be replaced at frequent intervals. In shopping complexes, the structure may be 'permanent', while the shop fitting will be frequently changed to keep up with modern fashions. This implies different selection criteria for different products within the same building.

One factor that is constant in our research findings is time. Specifiers would appear to be severely limited by time constraints. This is particularly true of designers working on refurbishment and rehabilitation projects where the choice of product cannot be determined until the work is opened up. Here, the pressure to make a decision in limited time is vital. Tools that help designers to access reliable information quickly, so that they can make informed decisions, would be welcome. However, even with the growth of information technologies and the trend towards electronic gateways, where filtered information is passed onto design organisations on a 'pay-to-use' basis, there is still no source of comparative data to assist the specifier. Indeed, it is questionable as to whether there will ever be such a directory.

There are a number of reasons for this. First is the sheer enormity of the task; the number of building products available means that there may be too many to compare. Reducing the number of products compared and listed in a directory would make the task easier to achieve, but then it is questionable as to how much use the directory would be to specifiers, simply because the product they may wish to use is not listed. Second is the problem of keeping the directory up to date. New products and minor product improvements are continually introduced, and thus the information needs to be monitored and revised frequently to maintain its currency. While the information can be updated quickly in electronic format, the actual process of listing the products characteristics against established benchmarks would be time-consuming. Third is the problem of checking all of the claims made by individual manufacturers in their promotional and technical literature. A trusted, independent, organisation would be needed if the information were to be trusted by specifiers, and someone has to pay for this. Fourth is that potential users of such an information source will have specific queries relating to their own particular problem. Such a directory might help them to narrow alternatives, but they will still need to access, and perhaps communicate with, the manufacturer in order to resolve specific queries.

Different levels of quality

One could be forgiven for thinking that the best product must be selected every time. The research findings reported earlier clearly show that the product's characteristics are only one of a number of issues considered by the specifier. With concerns over quality and liability, the focus is usually on the product perceived to be most durable and less likely to cause the specifier problems in the future. Indeed, with the ever-present threat of legal action against design organisations, the natural tendency is to 'over-specify' as a means of protection. Thus, buildings may be more expensive to build than need be the case; indeed, the lack of comparative data discussed above may well exasperate the situation. Questions have to be asked about the quality required by the client for a particular project and the quality actually specified.

Recently, an attempt has been made to try and establish an information system based on different quality levels. Specialist contractor procurement files (S-files) are being developed and will be piloted on specifiers as a means of aiding the designer and promoting better practice (Dixon, 2001). These files aim to provide generic specifications for a product type based on the collective experience of manufacturers, installers and specifiers, providing three different specifications to suit different quality levels.

Establishing and maintaining best practice

There is a very clear need for improving awareness about the importance of specifying buildings correctly, both for individuals and for design organisations. There are two inter-related areas to consider here, professional updating and reflective practice, key elements to the ideal of life-long learning.

The more efficient the specifying process, the better it is carried out, and the more time and resources can be spent on other (more profitable) tasks. Professional updating via continuing professional development initiatives and in-house training programmes can help to raise awareness of issues and help to refine and improve the manner in which tasks are carried out. Some of the larger design organisations organise their own in-house updating sessions for their staff. Smaller organisations have to rely on informal networking arrangements or on formal continuing professional development sessions organised by a professional institution and/or a commercial enterprise. More recently, the introduction of the vocational qualification in Architectural Technology at Level 4 (NVQ) may help to improve specification practice because it contains an entire unit dedicated to specification writing and decision-making.

Linked to continual professional updating is the philosophy of reflective practice; the ability to reflect on one's actions in the workplace with the aim of doing it better the next time. The argument for reflective practice is well rehearsed and its effectiveness clearly demonstrated in the workplace. However, time must be found to engage in professional updating and reflective activities and programmed into the working pattern of the design office.

There are three underlying and topical issues to consider.

Environmentally friendly products

A recently introduced aspect of the selection process is that of sustainability, the implication being that products should be selected that consume less energy in their manufacture or are in some other way less environmentally damaging, as discussed earlier. Because this is a relatively new criterion, we may assume that while there may be some products that have always been satisfactory in this regard, there will be many more that are not so. Thus, if specifiers are behaving in a responsible way, we would expect to see a shift away from some of the older products towards newer products that satisfy the new criteria. (Here, of course, we are ignoring the possibility of improvements within familiar products.) In other words, this should be a period of greater innovativeness among specifiers as they search for these new products.

We have no evidence that this is so. Indeed, there is every indication that it might not be so. While the adoption of these new products may have advantages for the planet, there is no indication that they have any advantage for either the specifier or the individual client. The reverse may be true. The adoption of such products may increase the risk of failure, or simply be perceived to do so. Social conscience does not necessarily overcome the tendency to continue to use familiar products. This is a rapidly developing area with a number of decision-making tools coming to market to assist the designer and specifier. Clearly, it is an area that design offices and individual specifiers need to give some thought. Consideration should be given to:

- the conservation and re-use materials;
- the conservation of energy;
- the minimalisation of waste;
- the minimalisation of pollution.

On an individual level, some specifiers have always pursued a policy of trying to specify products that have been manufactured locally, but not just because of concerns for the environment. The reasons for using local suppliers are:

1 they can get to site quickly if there is a problem during the buildings assembly; and
2 it goes some way to supporting the local economy, and it reduces transportation costs (and, of course, reduces pollution from unnecessary transportation).

Whilst such a policy is admirable, it is not always feasible. Clearly, the specifier has a duty to specify the product that best suits the client's particular requirements at the time.

Liability

This brings us naturally to the question of professional liability that appears to be an important influence on the behaviour of individual specifiers and their offices. Such considerations hinder the process of awareness through the gatekeeping mechanisms that are employed. Together with the pressures of time on the designer, they reinforce the tendency to adhere to the palette of favourite products. The dual considerations of liability and time are the most important factors in ensuring that specifiers are not innovators, for all their self-image that they are. Of course, it is in the interests of both building manufacturers and the publishers of architects' journals to reinforce this self-image.

An obvious way to reduce liability is to use products that have some kind of guarantee. As these have little worth if the company providing the guarantee ceases trading, the tendency is to use long-established companies with good track records. Another means of limiting liability is to use standard details supplied by the manufacturer. Companies who can offer specifiers these ways of limiting their risk are more likely to be able to market new products successfully. It is in a manufacturer's interest to present its product in a way that reduces both the perceived risk and the design time that the specifier will have to invest in adopting his product. Competent and readily available technical support appears to be far more effective than the employment of salesmen. However, good technical support, if not immediately available, may be useless. Again, it is a question of timing.

Small manufacturers may be unable to provide an adequate level of technical staff, but this is sometimes done through trade associations that may in effect be marketing organisations. Of course, large manufacturers also use such associations in a similar capacity, and these are often important in providing much-needed technical information. Yeomans (1988) has noted the importance of the International Trussed Plate Association in the adoption of trussed rafter roofs – in this case, acting to prevent their rejection by the industry. However, we have not been able to explore the extent to which information provided by such trade associations could reduce the design time and perceived risk of the product.

Office management

It is clearly in the manufacturer's interest to encourage innovative behaviour. This does not simply mean to encourage a specifier to use the manufacturer's latest product development but to use any of his products that may be new to the specifier. We may then ask whether it is in the interest of the office to encourage innovative behaviour. The answer is clearly that it is not, because of the penalties attached. Of course, there are those occasions in which the specifier has to be innovative, and the issue is whether offices can or should be managed so that the adoption of innovations is facilitated in such circumstances.

In offices of sufficient size, or for sufficiently large projects, it may be possible to employ someone whose role is to be aware of the full range of products available and the sources of information pertaining to them, the design manager or the chief specification writer. However, this will not be possible in the majority of cases because of the small size of the design organisations, and, rather than looking for ways of facilitating the employment of a wider range of products, offices would be do better to consider the effectiveness of their gatekeeping activities. This should not be taken to imply that they should be tightened to restrict even more the flow of information into the office. Rather, they should be managed in a way that encourages the flow of information on products that satisfy the requirements suggested above, i.e. those that are accompanied by information sufficient to facilitate good detailing and the provision of adequate technical support.

Best practice

This brings us to 'best practice'. Best practice may be defined in standards and/or research reports, although in reality, it is often deemed to be the practices that best work for a particular design office. When talking to specifiers to gather insights for this book, it quickly became apparent that individuals hold very strong views on the subject of specification, their best practice differing considerably from their peers, often within the same office. Many specifiers were open and honest in their opinions and claimed that although they knew that rolling specifications from one project to the next was bad practice, it worked for their organisation. Such habits were seen to be widespread and effective for design offices. We beg to differ for the reasons outlined earlier. What this shows, however, is that what we view to be bad practice is seen by many to be the best way of specifying for their particular circumstances, reinforcing our plea for better education and professional updating in this area. Indeed, we hope that the contents of this book will help students and practitioners to consider and reconsider specification matters with a view to developing best practice.

Future directions

Throughout this book, we have tried to portray the specification process as a rich and rewarding subject. We have argued that specification deserves greater attention from both academics and practitioners if real improvements in quality and value are to be achieved. By way of an epilogue, it is useful to look, briefly, at future developments in this area.

Specification trends

Mention has already been made of the world-wide move towards a performance-based approach to building design. Whether or not practitioners have the desire to specify buildings and their components entirely through the use of performance specifications remains to be seen. The current argument in the research community is for their exclusive use, but practitioners appear to be happier with the prescriptive method, or a mixture of performance and prescriptive methods. Whatever method becomes most fashionable in the future, there is still a need to raise the standard of specification writing and the decision-making process that precedes it.

Information technology

IT in construction is developing rapidly, and there are a number of tools available that may assist the specifier with time-consuming tasks. Information searches and retrieval, specification writing, and the tracking of products from manufacture to assembly on site are all made easier and more efficient through the adoption of appropriate IT. Such tools will continue to evolve, and are welcome. However, we should not lose sight of the fact that IT is only a tool. Individuals still need to make the decisions, confirm them, and take responsibility for the consequences of their actions.

Design management

We have witnessed considerable growth of interest in practical design management techniques over the past decade, in part due to greater attention from academics, but largely in response to an ever-competitive marketplace for professional services. Necessity has brought about greater awareness of management techniques with (mostly) improved performance as design offices learn to creatively manage their knowledge assets. Better awareness of time constraints and careful programming can both assist the specifier and help to ensure that the office is working profitably. The better the design management control, the less exposed is the organisation to claims. Indeed, in our conversations with practitioners, it appears that placing specification in a design and information management context can help an organisation to improve its effectiveness and ability to compete.

Manufacturers' input

The trend for greater involvement and co-operation between the manufacturers of building products and those who do the specifying is welcome. However, more effort is required if the communication gap between manufacture and design is to be narrowed. Integration, partnering, trust and supply-chain management are fundamental to this aim. So, too, is a better understanding of both the customer's requirements and the supplier's manufacturing constraints.

Final words

On a final note, we need to remind ourselves that to specify buildings effectively and efficiently requires talented individuals who are able to synthesise and apply a wide range of knowledge in a creative manner. Specifiers, regardless of professional background, need to work in a consistent managerial framework and have access to the latest tools to help them achieve their tasks. Properly resourced and managed, the entire decision-making process that we know as specification is key to providing a professional service and good-quality buildings.

References

Akin, O. (1986) *Psychology of Architectural Design*, Pion, London.

Allinson, K. (1993) *The Wild Card of Design: A Perspective on Architecture in a Project Management Environment*, Butterworth Architecture, Oxford.

Allee, V. (1997) *The Knowledge Evolution*, Butterworth-Heinemann, Boston, MA.

Andrews, J. and Taylor, J. (1982) *Architecture: a Performing Art*, Lutterworth Press, Guildford, UK.

Antoniades, A. C. (1992) *Poetics of Architecture: Theory of Design*, Van Nostrand Reinhold, New York.

AIA (1988) *AIA Handbook*, American Institute of Architects, Washington, DC.

Bass, F. M. (1969) A new product growth model for consumer durables. *Management Science* **13**(5), 215–227.

Banham, R. (1969) *The Architecture of the Well-tempered Environment*, Architectural Press, London

Barbour Index (1993) *The Barbour Report 1993: The Changing Face of Specification in the UK Construction Industry*, Barbour Index plc, Windsor, UK.

Barbour Index (1994) *The Barbour Report 1994: Contractors' Influence on Product Decisions*, Barbour Index plc, Windsor, UK.

Barbour Index (1995) *The Barbour Report 1995: The Influence of Clients on Product Decisions*, Barbour Index plc, Windsor, UK.

Barbour Index (1996) *The Barbour Report 1996: Communicating with Construction Customers – A Guide for Building Product Manufacturers*, Barbour Index plc, Windsor, UK.

Benes, P. (1978) The Templeton 'Run' and the Pomfret 'Cluster': Patterns of Diffusion in Rural New England Meetinghouse Architecture, 1647–1822, *Old-time New England*, Vol. LXVIII, Nos 3–4, Winter–Spring.

Bowes, J. E. (1981) Japan's approach to an information society: a critical perspective. *Keio Communication Review* **2**, 39–49.

Bowley, M. (1960) *Innovations in Building Materials: An Economic Study*, Gerald Duckworth & Co. Ltd, London.

Bowley, M. (1966) *The British Building Industry: Four Studies in Response and Resistance to Change*, Cambridge University Press, Cambridge, UK.

Bowyer, J. (1985) *Practical Specification Writing: a Guide for Architects and Surveyors* (2nd edition), Hutchinson, London.

Bradbury, J. A. A. (1989) *Product Innovation: Idea to Exploitation*, John Wiley & Sons Ltd, Chichester, UK.

BS 4940 (1994) *Technical Information on Construction Products and Services*, British Standards Institution, London.

Brawne, M. (1992) *From Idea to Building: Issues in Architecture*, Butterworth-Heinemann, Oxford.

Brown, L. A. (1981) *Innovation Diffusion: A New Perspective*, Methuen, London.

Building Design (2000) RHWL Partnership job advertisement, October 6.

Bullivant, D. (1959) The Problem of Information before the Architectural Profession and the Building Industry, *The Architects' Journal*, April 2; 512.

Cassell, M. (1990) *Dig it, Burn It, Sell it! The Story of Ibstock Johnsen, 1825–1990*, Pencorp Books, London.

Cecil, R. (1986) *Professional Liability* (2nd edition), Architectural Press, London.

Chapell, D. & Willis, C. J. (1992) *The Architect in Practice* (7th edition), Blackwell Scientific Publications, London.

Chermayeff, S. (1933) New materials and new methods. *Journal of the Royal Institute of British Architects*, 23 December, 165–173.

Chisnell, P. M. (1995) *Consumer Behaviour* (3rd edition), McGraw Hill Book Company, London.

Chown, G.A. (1999) Requirements for durability and on-going performance in Canada's objective-based construction codes. In: Lacasse, M. A. and Vanier, D. (eds), *Durability of Building Materials and Components 8*, Vol. 2, pp. 1527–1536.

Coleman, J. S. (1966) *Medical Innovation: A Diffusion Study*, Bobbs-Merrill, New York.

Concise Oxford Dictionary of Current English (1990), Clarendon Press, Oxford.

Cornes, D. L. (1983) *Design Liability in the Construction Industry*, Granada Publishing, St. Albans, UK.

Cox, P. J. (1994) *Writing Specifications for Construction*, McGraw-Hill Book Company, London.

Crosbie, M. J. (1995) Why can't Jonny size a beam? *Progressive Architecture*, June 1995, 92–95.

CSI (1992) *Manual of Practice*, Construction Specifications Institute, Alexandria, VA.

Cuff, D. (1991) *Architecture: The Story of Practice*, The MIT Press, Cambridge, MA.

Davies, S. (1979) *The Diffusion of Process Innovations*, Cambridge University Press, Cambridge.

Druker, P. F. (1985) *Innovation and Entrepreneurship: Practice and Principles*, William Heinemann, London.

Dixon, K. (2001) Specialist contractor procurement files. In: Emmitt, S. (ed), *Detailing Design*, LMU, Leeds, UK.

Edmonds, G. (1996) Trade Literature and Technical Information. In: Nurcombe, V. J. (ed.), *Information Sources in Architecture and Construction* (2nd edition), Bowker Saur, London.

Egan, J. (1998) *Rethinking Construction*, DETR, London.

Ellis, R. and Cuff, D. (eds) (1989) *Architects' People*, Oxford University Press, Oxford.

Emmitt, S. (1997) The diffusion of innovations in the building industry. Ph.D. thesis, University of Manchester, Manchester, UK.

Emmitt, S. (1999) *Architectural Management in Practice: A Competitive Approach*, Longman, Harlow, UK.

Emmitt, S. (2000) 'Changing the habits of a lifetime': a critical perspective on sustainable building. In: Erkerlens, P. A., de Jonge, S. and van Vliet, A. A. (eds), *Beyond Sustainability: Balancing Between Best Practice and Utopia*, Eindhoven University of Technology, Netherlands, keynote paper K2.

Emmitt, S. (2001a) *Architectural Technology*, Blackwell Science, Oxford.

Emmitt, S. (2001b) Technological gatekeepers: the management of trade literature by design offices, *Engineering, Construction and Architectural Management*, **8**(1) February 2–8.

Foxall, G. R. (1994) Consumer initiators: both innovators and adaptors! In: Kirton M. (ed.), *Adaptors and Innovators: Styles of Creativity and Problem Solving* (revised edition), Routledge, New York.

Gatignon, H. and Robertson, T. S. (1991) A propositional inventory for new diffusion research. In: Kassarjian, H. H. and Robertson, T. S. (eds), *Perspectives in Consumer Behaviour* (4th edition), Prentice-Hall International (UK) Limited, London.

Gelder, J. (1995) *Specifying Buildings: a Guide to Best-practice*, NATSPEC Guide, Construction Information Systems Australia, Milsons Point, New South Wales.

Gilfillan, S. C. (1935) (1970 imprint) *The Sociology of Invention*, MIT Press, Cambridge, MA.

Goodey, J. and Matthew, K. (1971) *Architects and Information*, Research Paper 1, University of York, Institute of Advanced Architectural Studies, York, UK.

Grant, J. and Fox, F. (1992) Understanding the role of the designer in society. *Journal of Art and Design Education*, **11**(1), 77–78.

Greenberg, B. S. (1964) Person to person communication in the diffusion of news events. *Journalism Quarterly*, **41**, 489–494.

Gutman, R. (1988) *Architectural Practice: A Critical View*, Princeton Architectural Press, New York.

Hagerstrand, T. (1969) *Innovation Diffusion as a Spatial Process*, University of Chicago Press, Chicago, IL.

Harris, C. M. and Dajda, R. (1996) The Scale of Repeat Prescribing, *British Journal of General Practice*, November, **46**, 649–653.

Heath, T. (1984) *Method in Architecture*, John Wiley & Sons Ltd, Chichester, UK.

Higgin, G. & Jessop, N. (1965) *Communications in the Building industry: The Report of a Pilot Study*, Tavistock Publications, London.

HAPM (1991) *Component Life Manual*, Housing Association Property Mutual.

Holden, R. N. (1998) *Stott & Sons: Architects of the Lancashire Cotton Mill*, Lancaster, UK.

Hubbard, B. Jr. (1995) *A Theory for Practice – Architecture in Three Discourses*, The MIT Press, Cambridge, MA.

Hutchinson, M. (1993) The need to stick to the specification, *The Architects' Journal*, 20 October.

Hutchinson, M. (1995) Specification substitution: the new construction industry ill, *The Comparative Performance of Concrete Roof Tiles*, Redland Technologies, Surrey, UK.

Latham, M. (1994) *Constructing the Team*, CM2250, HMSO, London.

Lawson, B. (1994) *Design in Mind*, Butterworth Architecture, Oxford.

Layton, C. (1972) *Ten Innovations*, George Allen & Unwin Ltd, London.

Leatherbarrow, D. (1993) *The Roots of Architectural Invention*, Cambridge University Press, Cambridge.

Leonard-Barton, D. (1995) *Wellsprings of Knowledge: Building and Sustaining the Sources of Innovation*, Harvard Business School Press, Boston, MA.

Lewin, K. (1947) Frontiers in group dynamics II. Channels of group life; social planning and action research. *Human Relations*, **1**, 5–40.

Macey, F. W. (1930) *Specifications in Detail* (fourth edition, revised by Brooke, D. and Summerfield, J. W.), The Technical Press Ltd, London.

Mackinder, M. (1980) *The Selection and Specification of Building Materials and Components*, Research Paper 17, University of York Institute of Advanced Architectural Studies.

Mackinder, M. and Marvin, H. (1982) *Design Decision Making in Architectural Practice*, Research Paper 19, University of York Institute of Advanced Architectural Studies.

MacLeod, M. J. (1999) Teaching prescribing to medical students, *Medicine*, **27**(3), 29–30.

Mahajan, V. and Wind, Y. (eds) (1986) *Innovation Diffusion Models of New Product Acceptance*, Ballinger Publishing Company, Cambridge, MA.

Maister, D. (1993) *Managing the Professional Service Firm*, The Free Press, New York.

March, J. G. (1994) *A Primer on Decision Making: How Decisions Happen*, The Free Press, New York.

Marks, P. L. (1907) *The Principles of Architectural Design*, Swan Sonnenschein & Co, London.

Mercer, E. (1975) *English Vernacular Houses*, HMSO, London.

Midgley, D. F. (1977) *Innovation and New Product Marketing*, Croom Helm, London.

Moore, R. F. (1987) *Specification and Purchasing within Traditional Contracting*, Technical Information Service 82, CIOB, Ascot, UK.

Nason, J. and Golding, D. (1998) Approaching observation. In: Symon, G. and Cassell, C. (eds), *Qualitative Methods and Analysis in Organisational Research*, Sage Publications, London.

Nawar, G. and Zourtos, K. (1994) And the walls came tumbling, *SPECnews*, Oct (cited in Gelder 1995, p. 129).

Newell, A. and Simon, H. A. (1972) *Human Problem Solving*, Prentice-Hall, Englewood Cliffs, NJ.

Nielson, M. and Nielson, K. (1981) *Risks and Liabilities of Specifications in Reducing Risk and Liability Through Better Specifications and Inspections*, American Society of Civil Engineers, New York.

Oostra, M. (1999) Initiatives for product development in the building industry: the architect as product innovator. In: Emmitt, S. (ed.) *The Product Champions*, LMU, Leeds, UK.

Oxford Thesaurus (1991) Clarendon Press, Oxford.

Parker, J. E. S. (1978) *The Economics of Innovation: The National and Multinational Enterprise in Technological Change*, Longman, London.

Patterson, T. L. (1994) *Frank Lloyd Wright and the Meaning of Materials*, Van Nostrand Reinhold, New York.

Pawley, M. (1990) *Theory and Design in the Second Machine Age*, Basil Blackwell Ltd, Oxford.

Peters (1988) Post medieval roof trusses in some Staffordshire farm buildings, *Vernacular Architecture*, **19**, 24–31.

Pool, I. de S. (1983) Tracking the flow of information, *Science* **221**, 609–613.

Potter, N. (1989) *What is a Designer: Things. Places. Messages* (3rd edition), Hyphen Press, London.

RIBA (1991) *Architects Handbook of Practice Management*, RIBA, London.

Rogers, E. M. (1962) *Diffusion of Innovations*, The Free Press of Glencoe, New York.

Rogers, E. M. (1983) *Diffusion of Innovations* (3rd edition), The Free Press, New York.

Rogers, E. M. (1986) *Communication Technology: The New Media in Society*, The Free Press, New York.

Rogers, E. M. (1995) *Diffusion of Innovations* (4th edition), The Free Press, New York.

Rogers, E. M. and Shoemaker, F. F. (1971) *Communication of Innovations: A Cross-*

cultural Approach (2nd edition), The Free Press, New York.

Rogers, E. M. and Kincaid, D. L. (1981) *Communication Networks: Toward a New Paradigm for Research*, The Free Press, New York.

Rosen, M. (1991) Coming to terms with the field: understanding and doing organisational ethnography, *Journal of Management Studies*, **28**(1), 1–24.

Rowe, P. G. (1987) *Design Thinking*, The MIT Press, Cambridge, MA.

Sabbagh, K. (1989) *Skyscraper: The Making of a Building*, Macmillan, London.

Saint, A. (1987) *Towards a Social Architecture: The Role of School Building in Post-war England*, London.

Sharp, D. (1991) *The Business of Architectural Practice* (2nd edition), BSP Professional Books, Oxford.

Shoemaker, P. J. (1991) *Communication Concepts 3: Gatekeeping*, Sage Publications, Beverley Hills, CA.

Simon, H.A. (1969) *Sciences of the Artificial*, MIT Press, Cambridge, MA.

Stone, P. A. (1966) *Building Economy – Design, Production and Organisation*, Pergamon Press, Oxford.

Symes, M., Eley, J. and Seidel, A. D. (1995), *Architects and Their Practices: A Changing Profession*, Butterworth Architecture, Oxford.

Tarde, G. (1903) *The Laws of Imitation* (Trans. Clews Parsons, E.) (cited in Rogers, E.M., 1995), Holt, New York.

Taylor, G. D. (1994) *Materials in Construction*, Longman Scientific and Technical, Harlow, UK.

Thornley, D. G. (1963) Design method in architectural education, Jones & Thornley (eds), *Conference on Design Methods*, Pergamon, Oxford.

Utterback, J. M. (1994) *Mastering the Dynamics of Innovation*, Harvard Business School Press, Boston, MA.

Wade, J. W. (1977) *Architecture, Problems, & Purposes: Architectural Design as a Basic Problem-solving Process*, John Wiley & Sons, New York.

Walton Markham Associates Ltd (1981) Communicating and selling to architects, *Architects' Journal*, August, 380.

White, D. M. (1950) The 'Gatekeeper': a case study in the selection of news. *Journalism Quarterly*, **27**, 383–390.

Willis, C. J. and Willis, J. A. (1991) *Specification Writing for Architects and Surveyors (tenth edition)*, BSP Professional Books, Oxford.

Windahl, S., Signitzer, B. with Olson, J. T. (1992) *Using Communication Theory*, Sage Publications, London.

WHO (1995) *Guide to Good Prescribing*, World Health Organisation, Geneva.

Wyatt, D. P. and Emmitt, S. (2000) The reference document template in whole life portfolio management, *Sustainable Product Information*, Proceedings of UICB/W102, Helsinki, Finland.

Yeomans, D. T. (1982) Monitoring design processes. In: Evans, B. et al. (eds), *Changing Design*, John Wiley & Sons, Chichester,

Yeomans, D. T. (1988) The introduction of the trussed roof rafter in Britain, *Structural Safety*, **5**, 149–153.

Yeomans, D. T. (1992) *The Trussed Roof: Its History and Development*, Scolar Press, Aldershot, UK.

Yeomans, D. T. (1996) Concrete mix design: putting theory into practice. In: Emmitt, S. (ed.), *Detail Design in Architecture*, BRC, Northampton, pp. 126–135.

UK.Yeomans, D. T. (1997) *Construction Since 1900: Materials*, Batsford, London.

Appendix

Postal questionnaire results and commentary

The results from the postal questionnaire are presented below with a commentary on the response to each question. This commentary is additional to the discussion of results presented in Chapter 7, in which the results are discussed against the Rogers model. As discussed in the main text, the postal questionnaire was useful in highlighting some of the important issues to be addressed by the observational research. The commentary is presented in italics to distinguish it from the results of the postal questionnaire.

Response

A total of 453 questionnaires were issued, of which 138 questionnaires were returned, giving a respectable response rate of 30.5 per cent.

Section 1

Job description:
Architect 118; technician 07; other/unknown 13

Sample age:
under 25 04; 25–34 13; 35–44 45; 45–54 42; 55+ 33; Unknown 1

This compares with other statistics presented by the RIBA and in Symes et al. *(1995), so the responses can be taken to be representative of a larger population of architects.*

Office size (by number of technical staff recorded at the respondents office):

	1–5 Staff	6–10 staff	11+ staff
Mackinder's sample by office size	2	4	18
RIBA (1991)	70%	15%	15%
Postal questionnaire by office size	64%	17%	19%

Margaret Mackinder visited 36 offices, of which 24 were private architectural practices, with the remainder drawn from local government offices and architectural departments of large companies. Her sample of private architectural practices (1980: 100–101) were larger offices than those of the postal questionnaire respondents. The postal questionnaire respondents are close to the RIBA (1991) figures and are more representative than Mackinder's sample.

Are you responsible for running jobs?
Yes 134 (98%) No 4 (2%)
Average number of jobs worked on in the past 5 years: 87.5 or 17 per annum (109 responses).

The four not responsible for running jobs were those who were under 25 years old.

Project type:
Commercial 122; Residential 115; Industrial 103; Retail 74; Leisure 73; Medical 61; other 16.

All respondents ticked at least two areas of specialisation, the majority indicated three different types of project worked on in the past 5 years, and others indicated four or five. Thus, all respondents had claimed experience of at least two different types of project, which is consistent with other research (e.g. Symes et al.*, 1995).*

Section 2

Q1. Listed below are some of the most popular journals: would you indicate which, in order of preference, you read.

	1st	2nd	3rd	4th	5th	6th	Total
Building Design	63	30	22	6	3	1	125
Architects' Journal	50	40	11	3	5	1	110
What's New in Building (p)	6	2	11	15	18	11	63
Building Products (p)	4	7	11	15	11	14	62

	1st	2nd	3rd	4th	5th	6th	Total
Architecture Today	5	15	16	14	4	5	59
Architectural Review	9	16	17	7	3	1	53
Building	6	10	14	8	8	3	49
Building Refurbishment	1	5	6	10	5	16	43
ABC & D (p)	2	5	2	7	10	6	32
RIBA Journal (unprompted)	10	5	4	30	1	2	25
New Builder	2	2	7	3	4	4	22
Blueprint	4	1	2	2	4	3	16
Other	3	2	4	4	–	2	15

(p) designates product journal.

Mackinder noted the importance of journals in staying up to date with manufacturers and products, but she was not specific about the type of journal and offered no statistical evidence. The purpose of this question, therefore, was to get an indication of the type of journal read by preference. Building Design *and the* Architects' Journal *were the most popular. The two product journals,* What's New in Building *and* Building Products *were the next most popular when all preferences were added together. However, they were mostly recorded as 3rd, 4th, 5th or 6th choice, which suggests that they are looked at less frequently than the journals that contain more news and less product advertising.*

This is a special communication channel from which specifiers may become aware of building-product innovations and part of the Rogers model. At the outset of the research it was felt that it was important to try and assess the importance of the product journals (which carry information about new building products) in relation to the professional journals (which carry some advertising). However, there is no simple direct way of testing the extent to which products are noticed.

Q2. In your opinion, do you feel that the journals you read influence your design decisions or influence your selection of materials/products?

Yes	32	23%
Probably	73	53%
Probably not	19	14%
No	14	10%

Mackinder's sample read journals to provide a general overview of the products available. The postal questionnaire respondents clearly felt that the journals did influence their design decisions or selection of materials/products, thus emphasising the importance of the journal (the specialist communication channel), although this was not supported by the subsequent observations.

Q3. Would you consider selecting a material/product on the strength of an advertisement or technical article in a journal?

Yes	9	7%
Probably	48	35%
Probably not	38	27%
No	43	31%

'No, but instigate checking on it'; 'Not advert alone'; 'Probably – select for further research before use'; 'No, further research needed and Agrément Certificate to be examined'; 'No, not solely'.

This question was designed to assess the function of advertisements. There was a slight tendency towards the negative, whilst qualifying comments confirmed that additional information was required. This tends to support Rogers' model of a search for knowledge following initial awareness.

Q4. Do you consult trade literature on a regular basis?

Yes	127	92%
No	9	7%
Unanswered	2	1%

A high number confirmed that they consulted trade literature on a regular basis, which supported Mackinder's work. However, it was not possible to ascertain whether this literature was held in a personal file of information (see Q5 below) or was from another source.

Q5. Do you keep your own file of product information?

Yes	111	80%
No	26	19%
Unanswered	1	1%

'Yes, or technical library'; 'Yes, basic information'.

A high proportion kept their own collection of literature. This supported Mackinder's findings where architects had a 'strong tendency' to develop a file of favourite products. The response of 80 per cent was higher than the Walton Markham telephone survey, in which 60 per cent of their sample said they kept a personal file of literature. There is a problem here because the trade literature consulted (Q4) may be that kept in the respondent's own collection. The two comments received indicated that the personal collection of literature may be relatively comprehensive in that it may contain technical information from sources other than from manufacturers to a collection of 'basic information', presumably the products used most frequently. The personal collection of literature is important because it is specific to the specifier and not part of the Rogers model, but the manner in which literature gets to be included in the personal collection and the influence of it in terms of consideration of building-product innovations can only be addressed by observational research.

Q6. When you receive product information from a manufacturer, do you; (a) expect to be able to make a full and detailed specification on the strength of it (b) call a representative to assist with the specification?

	(a)	(b)			
		Yes	Sometimes	Rarely	No
Yes	30	5	19	4	2
Generally	57	13	30	13	1
Generally Not	39	11	26	2	–
No	10	5	4	1	–
Totals		34 (25%)	79 (57%)	19 (14%)	3 (2%)
(unanswered 2)					

Note. The answer to part (a) and part (b) have been combined. For example, of those respondents that expected to be able to make a full and detailed specification on the strength of information from the manufacturer (they ticked the Yes box), five of them called a representative to assist with the specification, 19 called a representative sometimes, four rarely, and two did not.

(a) 'Yes, vital'; 'Expect yes, more often than not though it cannot be done', 'Usually requires verbal contact'.

(b) 'Yes, vital'; 'Yes, often have to, unfortunately; literature often inadequate'; 'Representative requested to assist if product unfamiliar, i.e. "Wonderproduct"'.

This question arose out of Mackinder's work and was designed to assess the purpose of trade literature, i.e. was it something to specify from, or more of a tool to get the specifier to contact the manufacturer, thus triggering a visit from the trade representative? Some of Mackinder's sample required 'basic information', whilst others expected a sample specification and detailed drawings to be included in the literature. Her sample questioned the quality of the literature, which was reinforced by the comments received in the postal questionnaire. In this survey, even those who expected to be able to make a full and detailed specification on the strength of the literature also telephoned the representative to assist with the specification, which tends to support Rogers' model of the change agent as an agent of reinforcement.

Q7. When selecting materials, do you decide:

	Always	Often	Rarely	Never
Precise trade name/material	19	98	16	–
Use a generic description	3	80	38	1

(Unanswered 2)

Respondents could and did tick both boxes recording a preference for using precise trade names of products over a generic description. The use of precise trade name and generic terms concurrently supported Mackinder's findings.

Section 3

Q8. Which of the following contractual arrangements have you used in the last 5 years?

	1st	2nd	3rd	4th	5th	6th	Total
Traditional contracts	127	12	–	2	–	–	141
Design and Build	8	63	7	1	1	–	80
Project Management	–	7	9	2	–	–	24
Management Contracting	–	9	2	3	2	1	17
Construction Management	1	2	4	-	2	–	9
Joint Venture	–	–	5	1	1	–	7
British Property Federation	–	–	1	–	–	2	3
Other (ACA)	–	–	1	–	–	–	1

(Unanswered 4)

Section 3 was designed to gather information specific to the building industry and outside Rogers' general model. A shift away from traditional contracts might produce a change in specification pattern, but traditional contracts were clearly the most popular and used in preference to design and build. This was the same as Mackinder's sample, despite a difference of 12 years between the two pieces of research. Thus, comparisons between the postal questionnaire results and Mackinder's observations are valid since both are primarily concerned with traditional contracts.

Q9. Do you find that the type of contract influences the products/materials you select?

Yes	24	17%
Generally	32	23%
Generally not	40	29%
No	39	28%
Unanswered	3	3%

Yes/generally 40 per cent, generally not/no 57 per cent. There is not a sufficiently clear difference to be sure of actual behaviour here, especially since we are seeking respondents views, and they may well wish to believe that their behaviour is unaffected.

Q10. On your last project, did you change a material/product or component during the contract as a direct result of a request by the contractor?

Yes	70	51%
No	62	45%
Unsure	6	4%

Yes comments received:

Non-availability/delivery times	31
Cost	12
To suit programme	5
Design and Build contract	5
Contractors request/advise	5
To simplify construction	2

Other specific comments:
'His idea was better'. 'Co-operation'. 'Following discussions with design team and contractor'. 'Reductions in tender price offered and accepted by our QS, (product similar to the one specified)'. 'Was the absolute equivalent and benefited the contractor cost wise'. 'The contractor wanted to use MDF for the window bottoms instead of the softwood specified. I had no objection'. 'Change due to new product becoming available'. 'Superior product/material'. 'The material was the equivalent of the one specified. Naylor drainage for Hepworth'. 'On the design and build project, the exact material is

often influenced by the general contractor'. 'The alternative product proved to be equally suitable for the situation'. 'Expedience'. 'Product was equal to that originally specified'. 'At request of contractor was asked to use alternative of equal quality to that specified'. 'Often working with small contractors – wish to use materials/methods they are familiar with/easily obtainable in small quantities'.

No:
'Not requested'. (5) 'Not suitable'. 'Change nothing if possible'. 'Original spec/price had been incorporated in the bill'. 'Once specified and included in contract documents variations spell trouble'. 'The alternatives suggested by the contractor were not of suitable quality'. 'Too late to alter the design'.

Other:
'This quite often happens'.

Mackinder noted the 'widespread influence' of the contractor, although no figures were reported. This question was designed to collect some quantitative information (not available elsewhere) and qualitative information. The response is important since it confirms that the specifier's innovation-decision stage is more complex than Rogers' model and supports an additional stage during which a product specified by an architect may face discontinuance through the action of the contractor.

Of the yes comments received, 31 noted non-availability/delivery times, which is primarily a problem for the contractor, not the specifier (although it may be a problem for the architect's office if the contract programme is affected). This may be the case, or it could provide evidence of an excuse to change the product to benefit the contractor. Cost was also noted, more of a concern to the contractor rather than directly to the specifier.

Some of the comments received indicated that discussion about products with persons external to the architect's office took place. Some quantity surveyor involvement was noted at this stage also.

Q11. On your last five jobs, were any specific materials or components requested by any of the following?

	Often	Sometimes	Rarely	Never
Client request	36	75	23	2
Planners' request	17	66	30	17
Contractors' request	13	64	41	8

	Often	Sometimes	Rarely	Never
QS suggestion	7	46	46	26
Other	2	6	2	3

This question was designed for comparison with Mackinder's sample. In her sample, both the client and the contractor were seen to be an important influence on specification decisions (the previous question confirmed the influence of the contractor). Three-quarters of her sample left items to be specified by the quantity surveyor, and three-quarters of her sample also noted the influence of the planner. The postal questionnaire respondents recorded less influence of the quantity surveyor than Mackinder's sample (this may indicate a change in practice since her work was carried out, or it may be reflective of the larger offices in her sample); otherwise it supported her work. This is important since it is an aspect of adoption behaviour that is not comparable to Rogers' examples.

Q12. Does your practice have:

	Yes	No	Unsure
An approved list of materials/products?	43	91	3
A blacklist of materials/products?	47	80	7

'Different list for different clients'. 'Yes, through personal experience only'. 'Unofficially (blacklist)'. (2)

Mackinder looked at standard specifications but did not address approved or blacklisted materials. This question arose out of the group discussion and, again, is not part of the Rogers model. However, it is important since the use of such lists may prevent or encourage the use of certain products. The majority of respondents answered 'yes' to both or 'no' to both lists.

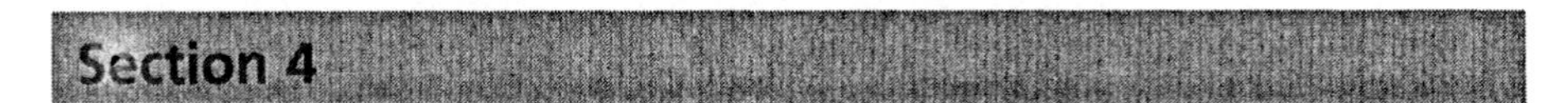

Section 4

Q13. Are you aware of price differentials between products with the same performance specification when you select them?

Always	12	8%
Often	60	44%
Occasionally	60	44%
Never	6	4%

'Quality comes first with pedigree'. 'Often – but suppliers and manufacturers often reluctant to make current prices available'.

Mackinder's sample reported cost as an important factor, but her sample complained that cost information was difficult to get since manufacturers did not make it available, a sentiment reported in the postal questionnaire.

Q14. Generally, when do you become aware of product cost?
(cumulative totals shown in brackets)

	Always	Often	Occasionally	Never
Outline proposals	7	39	45	21
Scheme design	10 (17)	58 (97)	38 (83)	8 (29)
Detail design/specification	40 (57)	65 (162)	16 (99)	1 (30)
Production information	39 (96)	50 (212)	18 (117)	2 (32)
Tender stage	48 (144)	38 (250)	20 (137)	4 (36)

'Varies depending upon size of project'. 'Ongoing general awareness with experience'.

Mackinder reported that cost awareness improved as jobs progressed and the quantity surveyor became more involved, which is supported in the information reported here.

Q15. Assuming you were aware of a range of product costs, do you feel it would influence your final selection?

Always	7	5%
Often	77	56%
Occasionally	50	36%
Never	4	3%

'No two products or sub-contracts are ever the same'.

The comment above helped to highlight the problems with asking such a general question and reinforced the need for some observational research. Most believe that cost has some effect, but we do not know how much. It may also depend upon the type of project and/or the type of client.

Section 5

Q16. Generally, how do you learn about new building techniques/methods and new products/materials?

	Always	Often	Occasionally	Never
Trade journals	21	84	18	–
Library	8	50	46	9
Direct mail	2	55	54	7
Trade representatives	2	51	62	5
Colleagues	8	46	58	9
Exhibitions	5	23	76	17

'None of these'. 'Direct mail always in the bin'.

The questions in this section were designed specifically to address the issues coming out of an early analysis of the Rogers model. This question was designed to assess how specifiers become aware of new products/building techniques, building-product innovations in Rogers' terms. It highlighted the role of trade journals (not supported by the subsequent observational research). Mackinder's sample varied in their readership of journals but viewed them as the most important source of information about new products, supported in the information reported here. Her sample noted the importance of colleagues exchanging views in the office, but this was largely in relation to products that had failed, not a source of information about new products.

The problem with this question is that it could not address whether this was passive awareness or whether the specifier searched for information about new products: active awareness. This is Rogers' chicken-and-egg problem, highlighting the need for some observational research.

Q17. How frequently do you select products that you have no previous experience of?

	u25	25–34	35–44	45–54	55+	?	Total
Often	–	–	–	–	–	–	– –
Occasionally	01	08	28	16	13	–	66 (48%)
Rarely	03	05	15	22	16	01	62 (45%)
Never	–	–	02	04	04	-	10 (7%)
	(04)	(13)	(45)	(42)	(33)	(1)	(138)

'Depends on definition of "no previous experience", and on type of product. (e.g. wall ties no, wallpaper yes)'.

The comment received indicated the problem of asking such a question and highlighted Mackinder's observation about the relative importance of products.

Results are shown against age group for this question and questions 18, 19, 20, 22 and 23. Although Rogers' earlier work stated that adopter age influenced the adopter's level of innovativeness, this has been played down in his later work. The problem here is that such an assessment could only be made against specific products, and therefore, only the total figures are used in the main text. The comments received for these questions are more significant and highlight the need for some observational research.

Q18. How long does a product have to be on the market before you would specify it?

	u25	25–34	35–44	45–54	55+	?	Total
Over 2 years	–	04	11	20	14	01	50 (36%)
1–2 years	01	07	15	05	09	–	37 (27%)
6–12 months	03	01	08	08	05	–	25 (18%)
Under 6 months	–	01	04	01	01	–	07 (5%)
Unanswered/depends	–	–	07	08	04	–	19 (14%)
	(04)	(13)	(45)	(42)	(33)	(01)	(138)

'Depends on type of product/product performance claims'. (6) 'Hypothetical, time of no concern'. (4) 'Depends on the type of product – a brick has little to prove'. 'Depends on client'. 'Well over 2 years'. '10 years'. 'The longer the better – particularly if completely innovative'. 'When it has a BBA Certificate'. '6–12 months if it was an alternative , say, type of floor vinyl, over 2 years for major items'. 'Depends on use intended, can't be answered, tend to use well-known products unless won't do job'. 'Depends on the producer'.

Time is an important factor in the Rogers model, but the comments received indicate the problem with asking such a question because it depends upon the importance of the product. Furthermore, the respondents are guessing how long a product has been on the market since they do not know (without checking with the manufacturer).

Q19. In the past year, have you specified products that are new to the market or new to you?

	u25	25–34	35–44	45–54	55+	Total
Yes	03	05	23	13	08	53 (38%)
No	01	08	22	28	25	84 (61%)
Unanswered	–	–	–	01	–	01 (1%)
	(04)	(13)	(45)	(42)	(33)	(138)

'Hardwood floors, Cumbrian slate, rough-cast render'. 'UPVC windows of a particular design, but from an established manufacturer'. 'Willan cavity tray specified before full production underway'. 'Glydevale roof vents by Willan'. 'Man made slates'. 'Roof coverings, cladding, polycarbonate glazing'. 'Onduline roof sheeting, (rigid corrugated board to substitute felt at shallow angles)'. 'Rockwool Rockclad'. 'Roof tiles, ventilation systems'. 'Artificial stone, kitchen units, steel doors'. 'Roofing materials'. 'Corovin roofing felt'. 'Formica Surell MFC'. 'Sandtoft Danum roofing tiles – new to me – never used them before'. 'Syphonic roof drainage'. 'Twinwall polycarbonate sheets'. 'Mainly to do with insulation (Celotex) and ventilation equipment and associated with recent Building Regulation update'. 'Broderic roofing; Tubular steel structures'. 'Flat roof insulation'. 'Varies, mostly finishes, e.g. floor coverings, lighting equipment'. 'Cladding materials'. (2) 'Ibstock Caplock'. (2) 'Belzona concrete repair and membranes'. 'Velfac windows'. 'Specialist equipment and aids for specific forms of disability'. 'Harling render'. 'TIBA cooker, stainless-steel posts in cavity walls'. 'Single membrane roofing materials'. 'Lighting components'. 'Alreflex 2L2 insulation'. 'Trench blocks'. 'T-Pren expansion joint by British Lead'. 'Curtain walling'. 'Roofing system our practice had not used previously although well used'. 'Products new to me too numerous to list'. 'Soffit ventilation and window ventilation'. 'Eternit Riverndale slates, DOW PR Roofmate'. 'Yes, generally 'low risk' interior design type products'. 'Fibre-reinforced concrete screed'. 'many', 'various'. (2), 'too detailed to answer'.

Thirty-two of Mackinder's sample of 36 offices (nine-tenths) preferred to use products that they had used before. Eleven (one-third) of her sample reported that it was office policy to avoid the use of anything new unless completely unavoidable. The respondents to the postal questionnaire appear to be more adventurous since 38 per cent claimed to have specified products in the past year that were new to the market or new to them. Of course, Mackinder's sample may have been obliged to use new products more than they would have liked. There is a difference in the question because the question used here needed to address actual behaviour rather than preferences.

The comments received show a wide range of products and once again demonstrate the need for some observational research based on specifiers' reaction to a number of specific product innovations.

Q20. If the manufacturer you normally use for a particular application does not produce your exact requirement, do you:

	u25	25–34	35–44	45–55	55+	?	Total
Attempt to find alternative (p)	03	08	35	29	25	01	101
Find alternative/Compromise	–	01	04	03	02	–	10
Ask to revise/find alternative	01	01	02	03	01	–	8
Ask to revise product (p)	–	01	01	02	02	–	6
Alternative/compromise/revise	–	02	01	03	–	–	6
Compromise and specify familiar (p)		–	01	01	03	–	5
Revise/compromise	–	–	–	01	–	–	1
Unanswered	–	–	01	–	–	–	1
	(04)	(13)	(45)	(42)	(33)	(01)	(138)

This question was designed to assess manufacturer loyalty (which may have formed a barrier to looking for building-product innovations). This was found to be low and was confirmed by subsequent observations.

Q21. The following products have been launched onto the market within the past 12 months: would you please indicate those you are aware of, those you have considered using and those actually specified.

Product	Aware	Considered	Used
Product 1	69	15	5
Product 2	66	17	3
Product 3	25	8	2
Product 4	31	3	2
Product 5	37	2	4
Product 6	19	0	0
Product 7	48	6	1
Product 8	24	13	1
Product 9	18	4	0

(20 of the 138 respondents indicated that they were unaware of any of the products.)

'I would point out that I would become aware of new products by reading trade literature or hearing about them, rarely a rep. calling to inform the office'. 'None'. (2). 'Never heard of any of them'.

This was designed as a cross-check to the diary of adoption and is discussed in Chapter 9. Please note that proprietary names were used on the postal questionnaire. These have been coded Products 1–9.

Q22. If you received details of an 'innovative' product, would you:

	u25	25–34	35–44	45–55	55+	?	Total
Wait until someone has specified (p)	02	05	17	22	19	–	65
Specify it on the next job (p)	–	04	12	09	02	–	27
Find out more	–	02	05	07	05	01	20
Unanswered	01	01	06	03	02	–	13
Dismiss it as too adventurous (p)	–	–	01	01	04	–	06
Wait for suitable opportunity	–	01	02	–	–	–	03
Depends	–	–	02	–	01	–	03
Invite representative	01	–	–	–	–	–	01
	(04)	(13)	(45)	(42)	(33)	(01)	(138)

'Investigate technical details if interested/explore possibilities'. (16) 'Specify it on the next job if appropriate/after thorough research'. (9). 'Wait until someone else has specified it and check its performance/wait to see results'. (3). 'Hypothetical – retain for consideration'. (2). 'Investigate the producer'. (2). 'Depends on the product'. (2). 'Request test results and certificates, BBA, BSI, TRADA, etc'. (2). 'Depends on job, use tried products unless "new" needed to do job'. 'Assess it myself, thoroughly and test it to destruction'. 'Specify it on the next job only if it was the best solution to the design problem in hand'. 'Depends if relevant – innovation does not prevent specification'.

The tendency to wait supported the views of Mackinder's sample and supported the opinions recorded in Q18. It also supports the Rogers model where only a small percentage of a social system are classified as innovators or early adopters. The comments suggest that the product would be investigated further if of interest, again supporting the Rogers model of initial awareness leading to a search for knowledge and the start of the innovation-decision process.

Q23 Have you ever specified products that you view to be novel?

	u25	25–34	35–44	45–55	55+	?	Total
Yes	01	02	15	12	11	01	42 (30%)
No	03	10	29	29	21	–	92 (67%)
Unanswered	–	01	01	01	01	–	04 (03%)
	(04)	(13)	(45)	(42)	(33)	(01)	(138)

'Trocal single-ply "Upside down" roof (specified in 1974)'. 'Yes at the time. After research decided to use single ply PVC membrane on roof construction, Trocal in 1979. Have had no problems whatsoever'. 'GRC cladding panels'. 'PVC rainwater goods, condensing boilers, combi boilers, heat pumps, man made slates etc.'. 'Roof boarding – Econoroof'. 'Stainless-steel slate hooks, 3D thermal economics sheeting'. 'Form of lead flashing'. 'Tufcote'. 'Canvas tenting as sun/rain shade to stadium'. 'Low water content boilers'. 'Fibreglass panelling externally'. 'Certain bathroom equipment, tapware and security entry devices'. 'GRP doors which have perfectly flush glass for clean rooms by Leader Flush'. 'Hewi ironmongery when it first came out in 1978'. 'Keps external walling system. Not used as client went bankrupt, but still hope to use in future'. 'Surrell, Corian'. 'Hepsleve'. 'Geoblock – plastic glass block by Cooper Clarke. Fermacell – gypsum bonded chipboard'. 'Curtain track lighting systems'. 'Metal imitation tiles'. 'Radway plastics dacatie cavity closer 1984'. 'Insulated cavity closer'. 'Polystyrene blocks as concrete formers'. 'Flat drainage pipes used as columns'. 'Monarfol roofing membranes'. 'Alreflex 2L2 insulation'. 'Items for detailing on buildings'. 'Cladding systems'. 'Very rarely'. 'Non-critical item'. 'Yes, too often'. 'Yes, can't remember'.

General comment received: 'Use of innovative product is often restricted, not by any doubt on performance, but by insurance companies indirectly and contractors unfamiliarity directly'.

Twenty-two out of Mackinder's sample of 36 (two-thirds) classed themselves as conservative in their approach to selecting materials (see also the comment on Q19). Sixty-seven per cent of the respondents to the postal questionnaire confirmed that they had never specified products that they viewed to be novel, thus supporting Mackinder's work. Care should be exercised in making a direct comparison here, but the answers to this question and those above indicated that approximately two-thirds of this sample act in what Mackinder's sample described as a conservative manner when selecting materials. This trait was confirmed by subsequent observation.

The 32 comments received from the 30 per cent who confirmed that they had specified products that they viewed to be novel provide a wide range of building products, ranging from large items such as cladding panels to much smaller products such as ironmongery. This list is important since it indicated that different respondents viewed different products as novel (no doubt based on different experiences), thus illustrating the difficulty of trying to assess the innovativeness of the sample. Furthermore, it both relies on memory recall and is not specific to, say, the last five jobs the respondent had worked on. Thus, there is no way of telling whether the respondents only did this once, or more frequently. Once again, it helped to demonstrate the need for some observational work.

Index

Page numbers in **bold** refer to figures.

Printed in the United Kingdom
by Lightning Source UK Ltd.
106387UKS00004B/14